Nanotechnology Applications and Markets

For a complete listing of related Artech House titles,
turn to the back of this book.

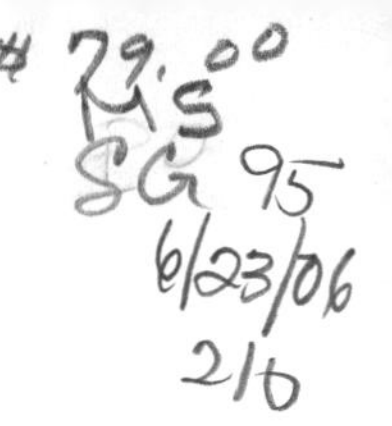

Nanotechnology Applications and Markets

Lawrence Gasman

ARTECH HOUSE
BOSTON | LONDON
artechhouse.com

Library of Congress Cataloging-in-Publication Data
A catalog record of this book is available from the U.S. Library of Congress.

British Library Cataloguing in Publication Data
Gasman, Lawrence
Nanotechnology applications and markets.—(Artech House nanotechnology library)
1. Nanostructured materials industry 2. Nanotechnology
I. Title
338.4'76205

ISBN 10: 1-59693-006-3
ISBN 13: 978-1-59693-006-3

Cover design by Yekaterina Ratner

685 Canton Street
Norwood, MA 02062

International Standard Book Number: 1-59693-006-3

10 9 8 7 6 5 4 3 2 1

To the victims of high-tech bubbles everywhere and everywhen and to Cynthia, my inspiration

Contents

Preface

> How very small the very great are.
> —*William Makepeace Thackeray*

> Small matters win great commendation.
> —*Francis Bacon*

Nanotechnology is, broadly speaking, an emerging technology that enables engineers to design and build new materials and products at the molecular level. The impact of nanotechnology is already being felt in the form of new computer memories that provide rapid access to stored data, that can hold more of this data than the minidrives used in iPODs and do not need any external power source to retain the data. It is being felt in the form of prototypes for photovoltaic cells that can literally be sprayed onto buildings or computers to provide cheap power sources. And it is being felt in the form of "nanoengineered" gels that speed the recovery of damaged nerve cells.

Not surprisingly, a technology this powerful is attracting attention. *The Economist*, *BusinessWeek*, and *Red Herring* (a magazine for venture capitalists) have all run cover stories or special reports on nanotechnology. New books on nanotechnology and the underlying nano*science* are also appearing at an accelerating pace. However, much of the coverage of the business aspects of nanotechnology in publications to date has been superficial at best. Most of the articles on nanotech that have appeared in the general business press have been written by journalists who do not specialize in this area, and therefore, the articles have consisted of a few generalizations and anecdotes with little or no serious attempts to point out where the opportunities in nanotechnology can actually be found. Furthermore, most of the books published to date are more about science and technology than economics.

All this is quite understandable. What is less understandable is that much of the coverage of nanotechnology business has a strong tendency to be either somewhat manic or somewhat depressive in nature. For instance, there are many articles that proclaim that nanotechnology will lead to the biggest economic boom since the rise of the microprocessor. These cite all the interesting work that is being done in this field and the major firms and adventurous start-ups that are involved. However, there are also many more articles that make the quite legitimate point that it is all too easy to overestimate the impact of nanotechnology and that most of the more spectacular expected products of the nanotech revolution lie two to ten years away, if not further.

In this environment, it seems that there is room for a book optimistic enough to characterize nanotechnology as a major upcoming business opportunity and realistic enough to recognize that it would take time to build businesses and develop products. Above all, a book that spent some effort going into the *specifics* as to the opportunities that nanotech brought, how long these opportunities were likely to take to develop, and outlined the appropriate business models necessary to make a profit from these opportunities. This book intends to provide this information.

However, this book is also a product of disaster and of lengthy memories.

The "disaster" in question was the fall of the telecommunications industry in the earliest part of this decade, which led me to refocus my research interests on nanotech. The "lengthy memories" were those of the management of Artech House, especially Mark Walsh. Mark commissioned a similar book from me on the telecom industry in 1986 and then, in 2004, asked me to write this one. I am looking forward to hearing from Mark again in 2022 with regard to writing a business-oriented book on picotechnology or even femtotechnology. Since I am currently in my fifties, I am not anticipating that Mark will call me once more in 2040, but if some of the more ambitious advocates of nanomedicine are to be believed that may indeed occur.

On Nanotech as an Epoch-Making Technology

There always seems to be a technology that defines the spirit of the times. I will call these "epoch-making technologies." If its strongest proponents are to be believed, nanotechnology is the next such technology.

At one time or another epoch-making technologies have included farming, the steam engine, electricity, automobiles, personal computing, and most recently, telecommunications. These technologies actually define society for a while, as, for example, when we refer to the "information age," or the "age of steam." These are the technologies on which the best engineers stake their careers and where the smartest businessmen put their money and entrepreneurial energies. Unfortunately, it is impossible to know for sure in advance whether a technology is really an epoch-making technology. When I was a teenager, there were few who would have argued against the notions that we would soon have electricity that was too cheap to meter as the result of atomic energy and that by early twenty-first century space travel would be as common as air travel. But somehow the space age and the atomic age have never actually occurred, at least not yet.

My guess is that many of you reading this book are doing so in the belief that (pun intended) nanotech is the next big thing, or, in other words, that nanotechnology is on the verge of becoming an epoch-making technology. If so, the examples of the space and atomic ages should serve as cautionary tales. Nonetheless, the not-so-hidden assumption behind the writing of this book is that, at the very least, nanotech is going to be behind the next surge in high-tech driven growth. It will not be too difficult to conclude from the chapters that follow that one of my motivations for writing this book is that I really do believe that nanotechnology will become an epoch-defining technology. This begs the question of why I believe this.

I think that there are essentially two answers to this question:

Nanotech Brings New Power The first of these answers is that, just as the wheel, the steam engine, and the computer did earlier, nanotechnology gives us a power previously unexperienced, that is, the power to consistently manipulate matter at the molecular or even the atomic level. As discussed in later chapters, we have always had this power to some extent, but the new tools that have emerged as a commercial reality in the last couple of years have raised our abilities in this regard by orders or magnitude (see Chapter 1). As a result, we are now in a position to create new materials, structures and devices to a degree never before possible. We are a very long way from the somewhat God-like powers that Eric Drexler talked about in the mid-1980s in his classic *Engines of Creation*, the book that first popularized the notions of nanotech. It is possible that we will never get to the stage that Drexler describes in his book, but it seems nearly certain that we will make it most of the way.

Nanotech Is in Tune with Today's "Megatrends" Nanotech is certainly not the only candidate for the next epoch-making technology. However, most of the other candidates seem to need nanotechnology to reach their full potential. For this reason, nanotechnology should be thought of as an enabling technology that in some way encompasses most of the other potential epoch-making technologies. This fact may confer on nanotech the status of a sort of "supermegatrend," and more reason to think of it as an epoch-making technology than the individual technology trends that it enables.

The role of nanotechnology as a broad-ranging, enabling technology is taken up again in Chapter 2, but it is also worth examining here, because it is so crucial to why many of us see nanotech as such a big business opportunity.

Mobile Communications Mobile communications is being touted by firms such as IBM, Intel, and Motorola as the next big thing in communications. These firms expect a major transition over the next decade to an environment variously called ubiquitous computing, pervasive computing, or invisible computing. In this brave new world of ultimate mobility, we will all carry multifunctional mobile computing/communications/entertainment devices that will be always on, taking in data from the Internet, sensor and RFID networks, and a plethora of other sources. Wired communications will remain in place for telecom infrastructure only, and wireless will be used for everything else. Mobile communications using the latest smartphones and notebook computers will transform the way that business is done and personal relationships are conducted.

According to Motorola, the most important limitation on the rise of ubiquitous computing is power. The current generation of lithium ion batteries work just fine for a cell phone used for the occasional short phone call. However, if used to power future smartphones, such a battery is likely to run down quite quickly. Nanotechnology comes into the picture here in a number of ways. It will help enable new kinds of power sources, such as better batteries, miniature fuel cells, and tiny photovoltaic panels, that will have greater power densities than today's batteries. It will also enable more energy efficient components and subassemblies for mobile devices. For example, a new generation of thin-film transistors built using organic molecules are enabling low-power plastic displays. Displays are typically the most power consuming subsystem in mobile computing or communications equipment. In addition to saving power, nanotechnology has the potential for bringing down the cost of mobile terminals and increasing the quality of visual output from these terminals.

Novel Energy Technologies With oil at record prices per barrel, with both India and China greatly increasing their consumption of oil, and the expectation that the West will grow even more dependent on OPEC for its existence, improved energy technology seems likely to be a focus of investment in the coming decade. This could mean technologies that improve the cost of production or the efficiency of use of fossil fuels. Or it could mean alternative energy sources, such as solar power, wind power, biomass, and the like. Either way, the energy industry and energy consumption patterns are likely to be quite different in 2025 than they are in 2005.

Nobody is expecting nanotechnology to provide a wholesale replacement for fossil fuels any time soon. However, what nanotech does promise are ways to make fossil fuels go further and to extract them more efficiently. Nanoengineered catalysts can be used to better extract oil, or turn oil into fuel for cars. At the same time, nanoengineering is leading to better fuel cells and photovoltaics, pushing these otherwise rather marginal alternative energy sources into new and bigger markets. Most dramatically, nanotech has the potential to create new ways to store and transport energy, which, in turn, will enable entirely new architectures for the power grid. In general terms, nanotechnology holds out the prospect of radically arranging the economics of bringing power to the right place at the right time.

Biomedicine and Pharmaceuticals If you want to bet on a sure thing, then a dramatic increase in demand for biomedical and healthcare products is what you ought to be betting on. Right now, the youngest baby boomers are in their forties and the oldest ones have just reached their sixtieth birthdays. The uncomfortable truth is that all of them are going to need more health care quite soon. It is not just the shear volume of boomers that will create this new demand, it is their expectations that old age will be both comfortable and unusually long-lived that will accelerate the demand for products in the health care and life sciences. This attitude is summed up in the subtitle of the very nanotech-oriented life extension book coauthored by the inventor and futurist Raymond Kurzweil, *Fantastic Voyage: Live Long Enough to Live Forever*.[1]

Nanotechnology is having an impact on numerous aspects of biomedicine and pharmaceuticals. In the pharmaceutical industry it is enabling drugs both to be discovered and to be delivered more effectively. Increasing the likelihood that new drugs will be discovered is inherently important to the pharmaceutical industry, which famously relies on "blockbuster drugs" for its existence. However, this is happening at a time when the ability of the industry to generate such blockbuster drugs appears to be on the wane. The potential importance of nanoengineered drug delivery systems can be easily understood by the apparent ability of nanoengineering to replace chemotherapy with an injection of specially prepared nanoparticles that kill cancer cells with minimal side effects for the patient. Nanotech is also improving medical imaging with improved diagnostic imaging techniques. Regenerative medicine is benefiting from gels that provide structure for nerve cells to grow back after injury, including improved stents for heart patients and even artificial blood cells.

Ubiquitous computing, a transformation of the economics of energy, and radically new pharmaceutical and regenerative medicine technology will all be epoch-defining in their way. What makes nanotech unique though is that it is an enabling technology that will make possible some of the most important developments in all of these areas.

A Cautionary Tale

As I shall discuss more in coming chapters, nanotechnology has much to do with a *growing* revolution in materials technology and, as such, it should be no surprise that it is an epoch-making technology in the manner defined earlier. After all, many of the great epochs of human history have been defined in terms of materials, such as stone and bronze, thousands of years ago, and more recently, steel and silicon. The silicon revolution, of course, was all about the ability of engineers to exploit the semiconductive properties of silicon to build microprocessors, computer memories, and other devices. This led directly to the most recent epoch-making technology that has generated important new business revenues: telecommunications.

Telecommunications[2] became the epoch-defining technology of the 1990s for a number of reasons, but primarily because an entirely new worldwide networking infrastructure had to be built to support the Internet. This infrastructure was digital, data-centric and, above all, optical and it replaced the old analog voice-centric—(and mostly copper)—network. During the period that telecom was in the ascendant, it was where the big money was to be made, whether you were a stock market investor, businessperson, or engineer. Individual investors rushed to get their money into the latest telecom IPO. Entrepreneurs and CEOs did all they could to promote their businesses as being telecom businesses. (Carl Russo, the head of Cisco's optical telecom business unit at the time, joked that if someone had opened "The Optical Bagel Store," he would have been mobbed by venture capitalists anxious to invest money.) Fiber-optics engineers who left college with the expectations of modest salaries in academe or some large industrial lab suddenly found themselves with more job offers than they could handle, many of them fully equipped with stock options potentially worth millions.

Within a few years, the telecom boom turned to bust as the digitalization of the network infrastructure (which had actually begun in the 1970s, but which went into overdrive with the arrival of the Internet) came to an end, as did the first rush of headlong growth from the Internet. The telecom industry went into a severe recession from which it is just now slowly recovering. In the process, the high-paying jobs and stock options all went away and fortunes were destroyed. Nonetheless, becoming part of an epoch-making technology wave has been rewarding for some. Many of the fortunes built during the telecom go-go years were invested wisely and preserved. While many of the upstart firms that appeared to take advantage of newly emergent opportunities in telecom have gone the way of all flesh, others have survived and are even beginning to flourish as the telecom industry bust gives way to telecom industry maturity.

The telecom experience should be a cautionary tale to anyone thinking of becoming involved with a nanotechnology business, whether that business is a group within a major corporation or a more entrepreneurial effort. There is a message here for investors, too. The three biggest lessons that to be learned from what happened to telecom are the following:

- *It doesn't take an avalanche.* The telecommunications industry provided profitable opportunities for firms for over 100 years before the telecom revolution. *So even if nanotech turns out to be something less than an epoch-defining, nanotech may still be responsible for creating new businesses, both large and small.*
- *All good things come to an end.* Everyone knew that the extraordinary growth that characterized telecom in the late 1990s would come to an end some day. (Though few believed the industry would crash so spectacularly.) It would not be the greatest surprise if, at some time in the next few years, nanotech also went through a similar boom-bust cycle. It is easy to imagine, how, for example, a high-profile IPO could spawn a nanotech boom.[3] In such a boom, many firms can be expected to overreach in terms of both their product offerings and their revenue expectations. In the telecom boom, several firms expected to build businesses with billion dollar revenues with huge optical switching products. In actual fact, the market demand for such switches never changed much during the boom and neither did the size of switches required. *If a nanotech boom occurs, it will be important for nanotech businesses not to be swayed by the hype and keep a firm grasp of market realities to build sustainable businesses. Almost all of the telecom boom era optical switching firms have now disappeared.*
- *Some things endure.* As I have already noted, there are firms that came out of the telecom mess fairly intact. In some cases, this is merely because they were large firms with an established base of even larger customers. However, some thought through the benefits of their

products and could present their stories to customers in a convincing way. The survivors' marketing stories have typically had to do primarily with reduction of capital and operational costs rather than the latest networking technology. *It will behoove emerging nanotech firms to focus on relatively easy-to-prove cost-related benefits than on the gee-whiz features of a revolutionary technology.*

On the Other Hand—Reasons for Nano Optimism

The experience of the telecommunications industry over the past few years provides plenty of reasons for nanotech firms to be quite cautious in their product and marketing strategies. Even if nanotechnology eventually turns out to be an epoch-making technology in the way defined earlier, the best that can probably be hoped for is that the nanotech epoch will last for a decade. In the course of history, technology defined epochs have tended to last for shorter and shorter periods. When nanotech gives way to something else, managers and engineers who had previously believed that they had high-flying careers at booming nanotech firms, and investors who saw their fortunes being made in nanotech, will suddenly find themselves a lot poorer or even unemployed, as we have seen happen in the telecom industry over the past few years.

However, the telecom analogy can also be taken too far. For one thing, many of the new products that appeared during the telecom boom era seemed to be way out of line with what the customers actually needed. The example of optical switches has already been given. It is probably too early to be sure of the general product directions of nanotechnology firms, but the fact that their products are so well tied in to the needs of some leading industry sector trends (enhanced mobile communications, better healthcare in an aging society, and lower energy costs, for example) is certainly a reason for optimism about the commercial future of nanotech.

This is a good argument to bring up to counter the comments of nano-naysayers. For, although there is plenty of hype about nanotech, there is also plenty of skepticism. This is healthy up to a point, as some of the pain of the telecom bust might have been avoided if there had been more thoughtful criticism during the boom. However, in my opinion, the high-tech bust of the earliest part of this decade and the related drop in equity values have now created a pessimist attitude in some quarters about *all* new technologies. This attitude regards all such technologies as somehow too risky or too far out to be taken seriously. There is no doubt that many of the applications of nanotech are "out there" in terms of timeframe, but others are not.

There is also no doubt that there are risks involved with any new technology. That is just the point. Most of the criticisms that can be levied against early stage nanotech, could be targeted towards *any* new technology. They could have been, and were, used against many new technologies of the past, including computing, networking, and steam engines. The individuals and firms who got into these fields early, and when the skeptics were ranting, often (although not invariably) did quite well financially.

A Personal Note

I first started thinking about the impact of nanotech back in the mid-1980s when I read Eric Drexler's seminal nanotech work, *Engines of Creation*.[4] Over the years, I kept referring to that book. When I did so, I wondered whether Drexler's vision could ever come to be. But I didn't wonder too long or too hard. The Drexlerian conception of nanotech seemed a long way off.

At the time, and then for almost another two decades, I earned my living by analyzing the commercial impact of new technology. But the technology I analyzed was telecommunications technology. My expertise lay in examining emerging communications technologies, matching their characteristics to market needs and predicting and quantifying where the market opportunities lay. This work sometimes took me close to the nanotech realm. At one point, I focused heavily on what new optical materials and developments in optical integration could bring to the market. I also carried out quite a few consulting assignments covering the potential for MEMS products in the telecom and data communications markets. (In those days, MEMS was supposed to be quite closely related to nanotech.) This enabled me to develop conceptual models relevant to the commercialization of nanotech. It left me with an overwhelming feeling that (and please forgive the pun) nanotech was going to be very, very big some day.

As we have seen, the telecom boom happened in the late 1990s, although the market had been growing rapidly and experiencing rapid technological change for almost a decade previously. Then the boom was swiftly followed by a telecom crash, the disaster mentioned earlier. As these words are being written, the telecommunications sector is showing signs of growth again, but, with the exception of developments in wireless, it is not being driven by technological innovation. Instead, the telephone companies are merely playing catch-up after a few years of failing to update their networks.

As I looked around for where genuinely novel technology was being developed, I discovered that much of the most exciting technology could now be considered as falling under the nanotechnology label. While nanotech has been around for quite a while in one form or another, it seemed that it has suddenly gained critical mass. The kinds of skill sets that we had developed at my old telecom industry analyst firm, CIR, now seemed like they could be easily applied to the budding nanotech market. With this in mind, my business partner, Robert Nolan, and I set up NanoMarkets LC (http://www.nanomarkets.net), which emerged out of long discussions that we had in 2003. As this book goes to press, NanoMarkets LC is well into its second year of providing focused market research and product planning in the form of consulting, newsletters, and relevant reports and other publications.

However, the kind of nanotechnology that provides the impetus for writing this book and is the focus of NanoMarkets work too is still a long way from the kind of nanotechnology I read about in *Engines of Creation* a couple of decades ago. Drexler's idea boiled down to a vision of nanotechnology as nothing less than molecular manufacturing. In this vision, we start with some kind of cheap feed stock and through a self-assembly process build high value materials and products. The kind of nanotechnology that I mostly will be talking about through most of this book is considerably less ambitious and encompasses a diverse group of areas, all of which have to do with engineering at the sub-100 nm level, this being the pervasive definition of nanotechnology these days.

Why This Book?

I shall have more to say about the Drexlerian vision and the more garden variety of nanotechnology in the main body of this book, because the debate between the Drexlerians and (what is becoming) more mainstream nanotechnologists is basic general knowledge for anyone who wishes to become part of the nanotechnology community. However, the details of that debate have only a marginal relevance to business opportunities in the nanotech space—the primary topic of this book. For those who have been caught up in all the hype surrounding nanotech, this book is intended to provide a gentle push towards reality. For those who understand that some of the pessimism about nanotech is also overstated and that many of the spoils of the nanotech revolution will go to the early entrants, this book is intended as a guide to how to make that early entrance.

Ultimately, this book is intended to answer two questions: Where are the business opportunities to be found in the nanotech, and when are they to be found? The book's objective is to look at nanotechnology the way that a businessperson would look at it. However, this should not be read to imply that the book will be of no interest to engineers, economists, investors, and academics.

On the contrary, it is the potential for business successes that is likely to be the ultimate fuel behind the interest in nanotech displayed by these other groups. The motivation behind the book is the strong sense that after decades examining the commercialization of advanced technologies, I can conclude that materials science-based business opportunities (that could loosely be subsumed under the rubric of nanotechnology) are on the verge following a sharp upward curve.

At the same time, having lived through the rise and fall of telecom, I have made one of this book's objectives to provide business with a framework to determine where genuine opportunities in nanotechnology lie. This is done in the belief that, to quote the highly successful venture capitalist, Vinod Khosla, "sometime in the next few years we will go through with nanotechnology the same kind of bubble we went through with the dot.com boom."[5] When that bubble occurs piles of dumb money are going to evaporate.

Organization of This Book

Chapter 1 is intended to set the scene for the balance of the book. I define nanotechnology and briefly review its history. I explain why it is that nanotech has now reached a critical mass in terms of commercialization to the point. I also delve into the tools that are becoming available to measure and manipulate matter at the nanoscale.

Chapter 2 is, in a sense, the core of this book. It discusses the thorny issue of whether nanotech can—or even should be—thought of as an industry in its own right and then goes on to examine the different categories of opportunity that nanotechnology presents to business development, marketing and product managers, and the different classes of products that are emerging from nanotech R&D. Also in this chapter I take a look at the relationship between nanotechnology and MEMS technology, two areas which some believe are quite close, and which together are sometimes considered under the heading "small tech." I continue in this chapter with a discussion of the special intellectual property and financial issues that are emerging in the nanotechnology space and conclude with an overview of the government programs around the world that have been designed to jump-start nanotech R&D.

The next three chapters each deal with new business opportunities for nanoengineered products in three industry groups that will be critical to the future of nanotech. It is just guess, but I suspect that more than 80 percent of nanotech business opportunities come from these industries, each of which face some kind of "crisis," that nanotech will be extremely helpful in making less severe.

The first of these industry groups considered in this book includes electronics and semiconductors and here nanotech is already helping to ameliorate the problems that are arising as Moore's Law moves the entire semiconductor industry into the nanosphere. It is also leading to brave new products such as sensors that can be sprayed onto surfaces and televisions that have the video quality of a CRT, but are thinner and lighter than any of the thinnest of today's flat panel displays. In Chapter 4 I consider the applications of nanotech to energy. As I have already noted the crisis in this sector is that the cost of the hydrocarbon-based fuels on which so much of our energy is based is going through the roof. Nanotech can help in numerous ways here. It can make petroleum extraction and use more efficient, provide various alternatives to hydrocarbons and offer entirely new ways to transport and store energy. In Chapter 5 I analyze the impact of nanotech on the healthcare and pharmaceutical sectors, which, again as I have already noted, face the crisis (and the opportunities) presented by populations of aging baby boomers, all of whom apparently are confident in their expectations to be healthy centenarians in a few decades.

In Chapter 6 I survey several other industry sectors that are likely to be impacted by nanotechnology and examine the opportunities in these sectors. Finally, in Chapter 7, I provide a framework for assessing the impact of nanotech on your company, which will hopefully lead to spotting new opportunities. Or at the very least, avoiding major difficulties.

Each chapter concludes with (1) a section titled "Key Takeaways from This Chapter," which summarizes three or four of the main points made in the chapter to which it is attached; and (2) a recommended reading list appropriate to the matters discussed in the chapter.

Acknowledgments

While I am entirely responsible for the opinions and other content found in this book, any merits that this book may have are due in no small part to the many people who I have talked with in the nanotech community over the past couple of years. These include the numerous executives and engineers at firms great and small with whom I have spoken as part of the regular briefings that I receive from key nanotech firms in my role as principal analyst with NanoMarkets, LC. I would also like to thank Howard Lovy and Paul Holister, who provided most of the content for the two reports that launched NanoMarkets and have been important sources for me of news, gossip, and understanding about nanotechnology ever since. Paul was the author of the first comprehensive study of nanotech commercialization ever published and Howard ran the best nanotech blog there is. Thanks are also due to Scott Mize, who formerly ran the Foresight Nanotech Institute and to Darrell Brookstein, who is the entrepreneur behind the www.nanotechnology Website. Both have been helpful in a number of ways.

I would also like to thank Rob Nolan, my business partner of almost a decade and cofounder of NanoMarkets, who has been the main source of marketing inspiration for the business. At Artech House, I would like to thank Mark Walsh, whose role was described in some detail at the beginning of this preface, as well as all the other editors that I have worked with there. Last, I would like to thank the members of my family, Cynthia, Antonia, and Andrew, without whom I would not be where I am, or who I am, today. Cynthia is the key to my success, Antonia the key to my staying youthful, and Andrew is the key to my exploring new worlds.

1

Nanotechnology: An Overview

> Nanotechnology is fast becoming as pervasive a cultural icon as TiVo or Levitra. The wizardry of building teeny things that are measured in one-billionths of a meter has begun to figure in Hollywood movies, in bestselling novels—even in Jay Leno's monologues.
>
> —*Fortune Magazine*[6]

Disputes and Definitions: What Is Nanotechnology?

The quote cited above conveys some of the excitement that currently surrounds nanotechnology. But what exactly *is* nanotechnology? Definitions of nanotechnology are easy to find, but hard to agree on.

The philosopher of science, Karl Popper, makes the point that definitions of any kind are merely conventions. We should judge them on how useful they are, not on how true they are.[7] Definitions are not, he says, something that we should take too seriously and certainly we should not fight about them.

In this book I will take Popper's advice, noting that such tolerance has not been characteristic of the history of nanotechnology so far. Even these days, in the earliest times of nanotechnology as a commercial activity, there have already been bitter disputes over what nanotechnology is and isn't. Is nanotech a synonym for molecular manufacturing? Or is it a broad term covering manufacturing processes that deal with very small features and devices? I will discuss these disputes, because anyone who becomes involved with the world of nanotechnology has to know about them. But I won't take them too seriously.

The purpose of this book is to point out new business opportunities. Consequently, its writer believes that the definition of the term "nanotechnology," used in the book, should be judged only on its usefulness in achieving this goal.

Some of the topics covered may thus alarm purists, who view nanotechnology as something very specific, that is, a technology in which one builds complex systems by rearranging matter at the molecular or atomic level. By contrast, I am going to be eclectic in my approach and will cover the broad range of topics that are typically discussed under the heading "nanotechnology," in the general business press, specialized publications, technical literature, and so on. On the other hand, I do not completely agree with some of the pragmatists in the nanotechnology community, who tend to dismiss writings on molecular engineering, nanomachinery, and the like, as belonging in the realm science fiction—and harmful science fiction at that.[8]

None of this, of course, means that we can do without a definition of nanotechnology altogether. A useful point to start, if only because it is so often quoted, is the official definition used by in the U.S. National Nanotechnology Initiative (NNI). The NNI says that nanotechnology, must involve all of the following:

1. Research and technology development at the atomic, molecular, or macromolecular levels, in the length scale of approximately 1 to 100 nanometer range.
2. Creating and using structures, devices, and systems that have novel properties and functions because of their small and/or intermediate size.
3. Ability to control or manipulate on the atomic scale.[9]

For the reader of this book, the truly important takeaway from the NNI definition is that nanotechnology involves the engineering of "structures, devices and systems that have novel properties and functions because of their small and/or intermediate size." Throughout this book we will be primarily concerned with the business opportunities that result from this kind of engineering.

A Very Brief Guide to the Science Behind Nanotech: Why 100 Nanometers?

The range of 1 to 100 nanometers mentioned in the NNI definition pops up continually in the nanotechnology literature. For example, one of the most popular introductions to nanotechnology defines nanoscience as "the study of the fundamental principles of molecules and structures with at least one dimension roughly between 1 and 100 nm."[10]

But just what is so special about 100 nm? Why not 10 nm? Or 1,000 nm?

The answer is that under the 100-nm level, engineers begin to deal with properties of materials that ordinary engineers can quietly forget about. In particular these include:

Quantum Mechanical Effects Most readers may be aware that atomic and subatomic particles are subject to the often bizarre laws of quantum mechanics. These laws enable a particle to (a) behave sometimes like a wave and sometimes like a true particle (complementarity), (b) be in two different places at the same time, and (c) effect the physical properties of another particle that may be millions of light years away (entanglement), to name some of the odder quantum mechanical phenomena. Quantum mechanics also establishes that it is impossible accurately measure both the location and the velocity of a particle simultaneously. Although some knowledge of quantum theory is necessary for any budding nanotechnologist, to discuss this topic in depth would go well beyond the objective of this book. There are many excellent books, and at every technical level, that can provide the reader with an introduction to quantum mechanics.[11]

Surface Science Effects Although quantum mechanical effects are the ones deemed most newsworthy, classical physical laws can also produce some very surprising effects in very "thin" nanomaterials; that is, materials where the surface-to-bulk ratio (the number of atoms bordering a surface divided by the total number of atoms) is very high. As with quantum mechanical effects, surface effects can lead to surprising chemical, optical, mechanical, magnetic, and optical properties in certain materials. For example, one book on nanoscience notes, "In aerospace and automotive applications . . . [nano] materials made from metal and oxides of silicon and germanium exhibit superplastic behavior, undergoing elongations from 100 to 1000 percent before failures."[12] As we discuss later in the chapter on energy, the high ratio of area to volume is also particularly important in making nano-enhanced catalysts perform much better than regular catalysts.

The somewhat unusual ways that matter behaves at the nano level can be a opportunity. (That is mostly what this book is about.) The practical advantages of using nanomaterials that result simply from the smallness of the particles from which they are constructed are already well developed in certain industries, with others just a few years from commercialization (see Table 1.1).

The opportunities associated with "quantum weirdness," are not so numerous—at least not yet. However, "quantum encryption," an apparently uncrackable form of encryption that makes use of the quantum phenomenon of

Table 1.1
Selected Opportunities for Nanomaterials

Industry	Materials/ Opportunity	Advantages of Nanomaterials	Important Future Directions and Opportunities
Aerospace	Nanomaterials and nanocoatings are being used for the bodies of aircraft and in aerospace components.	It has been claimed that the use of nanomaterials can increase the fatigue strength of aerospace materials by as much as 300 percent.[13] Nanomaterials may also be considerably lighter, reducing the fuel required—a critical issue in today's highly unprofitable airline industry. Nanomaterials may be especially useful for space vehicles that must meet extreme conditions—especially with regard to heat.	Materials with embedded nanosensors that constantly monitor the state or the airframe for safety and other concerns. Fuel additives/nanocatalysts that make the use of aircraft fuel more efficient.
Automotive	Nanocrystalline silicon nitride and silicon carbide have been used in springs, ball bearings, and other automotive components.	These materials demonstrate impressive mechanical and chemical properties that contribute to both the manufacturability and longevity of these components.	Coatings and materials for antifogging mirrors and windshields. Fuel additives/nanocatalysts that make the use of gasoline and diesel more efficient. High-resolution dashboard displays using organic electronics.
	Nanocrystalline ceramic liners for engine cylinders.	Zirconia and alumina liners have been used to retain heat in cylinders and improve the efficiency of combustion.	

Table 1.1 (Continued)

Batteries	The latest generation of batteries use nano-engineered aerogels[14] for separator plates.	These nano-engineered plates can store more energy than conventional plates.	Low-cost miniature hydrogen sensors for fuel cells. Improve fuel cells used nano-engineered membranes and catalysts. Lithium ion batteries that use carbon nanotubes or other materials to improve times between charge, energy density and especially the time taken to recharge.
Building materials	Aerogels for insulation and "smart windows" that darken when the sun is bright and get more transparent in dimmer light.	The structure of aerogels makes them excellent insulating materials.	Lighter materials that reduce construction costs and enable new kinds of architecture. Stronger materials with longer lifetimes that reduce construction costs and building depreciation. Biodegradable materials that reduce adverse environmental impact.
Machine tools	Nanocrystalline metal carbide materials for cutting and drilling. Nanoparticles for improved ceramics.	Nanocrystalline metal carbide materials provide harder, longer-lasting materials for drills and utting machinery. Conventional ceramics can be made less brittle and easier to work with through the addition of nanoparticles.[15]	Drills and cutting tools capable very hard materials and drilling especially small holes. May find use in semiconductor, MEMS and robotics sectors.
Televisions and monitors	Nanomaterials used to improve the resolution of CRTs. Carbon nanotubes used to create CRT-like field emission displays (FEDs). Organic polymer-based flexible displays.	Various zinc, cadmium, and lead nanomaterials have been proposed to produce smaller phosphors/pixels in CRT displays and hence better resolution. Carbon nanotubes make excellent emitters and prototypes of FEDs have been built that combine the visual quality of a CRT, yet may be only one inch thick.	Organic/flexible electronics will find applications in flexible PV and advanced lighting systems as well as in displays. Nanotube-based FEDs may find applications in X-rays, lithography, and lighting systems. FED displays may find a ready market for very large TV monitors and advertising displays.

Table 1.1 (Continued)

Regenerative Medicine	Nanoengineered gels and other materials are used to replace lost tissue or to provide structure for the regeneration of natural tissue. Current applications include bone replacement and nanostructures that help in the regrowth of nerves.	Nanomaterials are constructed at the size level of the human cell, which means that they are incorporated better into the body than other alternatives. For example, tissues can easily bond with nanoporous bone substitutes and nerve healing is improved when grown around nanostructures. Most nanomaterials used in such applications are also very strong, which has obvious advantages. However, there is some worry that the very fact that nanomaterials integrate well into natural body structures may cause body malfunctions, or even new diseases.	Future possibilities quickly stray into the area of science fiction, but nano-engineered heart valves and artificial kidneys realistic possibilities.

entanglement is already a reality and is being used by governments and in the financial services industry where its very high cost can be justified. Quantum entanglement is also the basis of quantum computing, which is still no more than just a research concept, but which promises computers more powerful than the most powerful supercomputer ever built. This power is based on an old idea that the performance of computers can be speeded up through simultaneously processing information through more than one computer. Parallel computing has been used in supercomputing applications for many years. A classical parallel computer simply grows its performance with the number of parallel processors used. Thus, a parallel computer with two processors of a particular kind is twice as fast as a regular computer based on one processor of that same kind. A quantum computer provides parallelism of a sufficient order of magnitude that it could perform in a matter of seconds, a problem that today's fastest parallel computer would take years to carry out.

I will discuss, in greater detail in later chapters, how quantum computers work and why nanotechnology will be needed to implement them. It is worth noting, however, that nanoscale effects do not always bring with them the tremendous opportunities set out above.

They can also be a nuisance. By way of a dramatic illustration of the kind of havoc that quantum events can cause, consider "soft fails" (aka "soft errors") caused by neutrons and alpha particles. Neutrons are tiny subatomic particles, whose whole behavior, and indeed, their whole existence, is shaped by the laws of quantum mechanics. However, computer chips have long been small enough to be seriously impacted by random bombardments of this kind of particle. As the name "soft fail" suggests, it is not that the chip itself is damaged in any way, but rather that a "bit flip" occurs. (A "0" becomes a "1" or vice versa.) This may not sound like a big deal, but Cypress Semiconductor reports that a single bit flip due to soft fail has caused hundreds of computers to crash and the billion-dollar factory of an automotive supplier to grind to a halt.[16] Furthermore, the problem is likely to increase as nanotechnology continues to impact the semiconductor industry. Cypress has also noted that as the density of SRAM memory chips increases, soft errors are likely to become a bigger problem.[17] IBM researchers have concluded that the "percentage of soft fails to hard fails becomes greater as the complexities of future chip technologies are increased,"[18] a trend that the advent of nanoelectronics will do much to accelerate.

To summarize, the opportunities and challenges associated with nanotechnologystem largely from effects that are very surprising when considered from the perspective of the everyday world that seems to obey the laws of classical physics. But this point is made in almost every book on nanotechnology. What is stated much less often is that nanoscience isn't always weird science. Nanoengineers also have to deal with all the properties of materials that ordinary engineers have to deal with, as for example, hardness, thermal properties, electrical conductivity, and flexibility.This is the significance of the lower bounds of the nano realm. Fall below that 1-nm range and we begin to enter the realm, not of nanotechnology, but of high-energy physics, where classical properties of materials are not an issue. To the extent that there currently exists a technology that impacts at this level, it is nuclear technology not nanotechnology.

The Early History of Nanotech

Most epoch-making technologies have been thought of, or even developed, in some primitive form, long before the epochs that they define. The steam engine, a technology that helped drive the industrial revolution in the 19th century, was originally described by Hero of Alexandria in 200 B.C.E.[19] As everyone knows,

Leonardo Da Vinci designed flying machines in the 15th century. The Wright Brothers flew the first heavier-than-air aircraft at the turn of the 20th century.

Nanotechnology also has its own prehistory of this kind. Medieval glassmakers sometimes used glass nanoparticles to provide color in stained glass windows.[20] This is real nanotechnology, since different size particles provided different colors. Depending on their size, gold nanoparticles can show up in glass as orange, purple, red, or greenish. However, just as Da Vinci's exploration of manned flight has little historical connection to the development of the aviation industry, so the nanotechnologists of the medieval era do not really have much to do with the current incarnation of nanotechnology, which can be dated back instead to a talk given by Richard Feynman.

Like Da Vinci, Feynman was a master of coming up with ideas that would not see commercial realization for decades to come. In addition to nanotechnology, Feynman thought up the idea for the quantum computer, mentioned above, an idea that is only now showing some signs of commercial potential. Unlike Da Vinci, however, Feynman's connection to today's state of the art is much more direct. As we shall see, there is a lot of debate about what exactly nanotech is and where it is headed commercially, but all the prominent individuals and various groups within the nanotech community seem ultimately to pay homage to Feynman and to his thought-provoking talk, "There's Plenty of Room at the Bottom." Back in 1959, when this talk was given,[21] the miniaturization of electronic circuitry was just beginning. (This would later lead to the creation of the semiconductor industry.) This trend raised the question for both engineers as well as for theoreticians as to how far we can miniaturize. Was there a size limit below which it would be impossible to do practical engineering, and not just in electronics, but in mechanical engineering also? What Feynman did really was simply to note that there was no good theoretical reason why engineering should not continue down to the molecular or even atomic level. In other words, he provided a sort of proof of the possibility of nanotechnology, although he did not use that term, since it had yet to be invented.

While Feynman is universally revered in the nanotechnology community, the same thing cannot be said about the other big name in nanotech, K Eric Drexler. Drexler was inspired by Feynman's ideas and took them to the next stage. He may have actually coined the phrase nanotechnology. He certainly was the first to bring it into common usage in his book, *Engines of Creation.*[22] Anyway, there can be little doubt that many of the people who are making careers in nanotech today first heard of it by reading Drexler's books (this was true in my case). However, for better or worse, Drexler has become a character about which much of the nanotech community has been polarized, although, as we note below, this has begun to change.

Eric Drexler's View of Nanotech as Molecular Manufacturing

Drexler views nanotech in a rather different light than many of the large corporations and entrepreneurs that are currently pursuing nanotech business opportunities. For Drexler, nanotech means primarily molecular engineering. Drexler does not deny that much of what passes for nanotech today is useful or potentially profitable, but he sees it as an evolutionary extension of older chemical engineering, rather than something that can change the world.

Drexler sees nanotechnology as something that can change the world the way that the steam engine changed the world. Molecular engineering can bring about a new "diamond age," the way that the steam engine brought about the industrial revolution.[23] Although neither Drexler nor anyone else expects to see the fruits of his kind of nanotechnology make an impact on the world in the next few years, the impact, if and when it comes, will supposedly be quite radical. Drexler envisions nano-scale self-replicating machines, called molecular assemblers, that could use a cheap chemical feedstock and by rearranging molecules in the feedstock, create valuable products such as petroleum, diamonds, or much complex systems. The labor needed for this kind of molecular engineering would obviously be minimal and Drexler believes that capital costs would be quite low too. And the molecular assemblers would be able to do wonderful things, such as toremove pollutants in the air, crawl around in our blood vessels and cure disease, and create extremely valuable products from abundant materials. As a article about Drexler in *Wired* magazine puts it, Drexler has "shining dreams of unprecedented material abundance, miracle medicine, and environmental revitalization."[24]

Drexler would say that what he is proposing is different from anything that has come before, envisioning an epoch-defining and one that it is certainly different from most of what now passes for nanotech,including most of what is discussed in this book. For this reason we will not discuss Drexler's more futuristic ideas about nanotechnology in any depth here. However, Drexler cannot and should not be ignored, not just because of his historical role in bringing nanotechnology into being as a unified discipline, but in his (albeit often indirect) influence on thinking about "real world" nanotechnology, nanotechnology in the laboratory, and about fundamental nanoscience.

For the purposes of this book there are really two takeaways to be had from Drexler's thinking and that of the community of scientists, technologists and business people who have been strongly influenced by his thought:

Nanotech Will Change the World (or Not) While, in some ways, the short-to-medium term opportunities that are created by nanotech are fairly mundane, the long-term commercial implications of nanotech may be truly mind boggling. The italics have been added to "may," because there are some critics of Drexler who say that Drexler is proposing isn't just something that is a long way off. It is something that is fundamentally impossible. This is not a book on nanoscience, so it is not the place to discuss in depth whether or not Drexler's self-replicating machines somehow run afoul of some fundamental laws of physics or chemistry. Those who want to explore this topic further should take a look at the debate between Drexler and the late Richard Smalley, a Nobel Prize winning nanotechnologist. Smalley says that Drexler's dream will never happen. Drexler counters that self-replicating machines can be built in principal, because living things are themselves an example of such a machine.[25]

Drexlerians Versus "Practical" Businesspeople Is a Dead Issue Drexler was a central figure in the a battle to get funding for molecular engineering provided as part of the U.S. government's National Nanotechnology Initiative. His side lost and, at the time, there was a certain amount of bitterness on the Drexlerian side, which believed that a major opportunity to boost American industry had been lost. Its opponents countered that their victory meant that scarce government funds would now be channeled into projects with a payoff measured in years rather than in decades. Whatever the truth of these assertions may be, there can be little doubt that the two sides have now come together to a considerable degree. The Foresight Nanotech Institute,[26] the think tank established by Drexler and others to support the Drexlerian agenda has broadened its definition of nanotech to include more than just molecular engineering. There is now considerable interaction, and certainly no animosity, between this group and Nanobusiness Alliance, which may be though of as the representative organization for "practical" businesspeople concerned with short- to medium-term business opportunities in nanotechnology.

Despite all the controversy over Drexler's work, his books will remain an inspiration to budding nanotechnologists everywhere, even though his vision is not likely to inspire many real world for-profit nanobusinesses in the near-term future. But that vision should not be dismissed completely, even by the most practical of businesspeople. In the laboratory at least, it is possible to create very simple molecular machines, which will be the building blocks for a future molecular manufacturing technology. It should also be noted that Drexler's vision of the world to come is not that far from what the current commercial nanotech is aiming at in at least one important sense. In addition to taking his insights from Feynman's seminal paper, Drexler also drew on molecular biology. What Drexler wanted is to make mechanical versions of biological subsystems such as ribosomes, and enzymes. This is not exactly what the current generation of nanotechnologists are aiming at, but what is important to them both is "biomimicry," which is looking to nature for ideas about how new materials and devices can be created.

Nanotech as the Evolution of Chemical Engineering

There is a whole other approach to nanotech and it is the one that currently dominates the commercial aspect of nanotech. It has little to do with molecular manufacturing and we have already gotten a flavor of it from the definitions explored earlier. Drexler would probably not even accept that it is truly nanotech. Indeed, having essentially given up the fight to use "nanotechnology" to denote something to do with molecular engineering, Drexler has apparently rechristened the approach that he advocates zeta-technology.[27] By contrast, Drexler would probably say that commercial nanotech today is actually an evolutionary extension of chemical engineering and this would be a fairly valid claim.

Organic chemists of the 19th century when attempting to create at the molecular level had available to them only what has become known as "shake and bake" methods. That is to say their methods were crude, consisting of adding reagents and/or heating until something useful occurred. Although this sounds hopelessly inefficient, even unscientific, it is important to realize that the shake and bake method could be very productive. It lead to the creation of the first synthetic plastics, dyes, and drugs, for example.

While shake and bake has not and will not be eliminated entirely, by the late 19th century chemists were beginning to develop a new methodology, that of rational synthesis. In this approach, one begins with a starting (or lead) molecule and converts it systematically to a new material with an understanding of chemical structure and the chemical processes that underlie that change. Rational synthesis is at the core of today's pharmacy and dates back to the 1860s when two German chemists synthesized alizarin (a dye substance originally taken from the madder plant). They did this by the planned modification of anthracene, an aromatic compound derived from coal tar.

Nanotools and Nanomanufacturing

The current breed of nanotechnology, in a way, represents the next stage of chemical engineering evolution after rational synthesis. While rational synthesis is better than trial and error (i.e., shake and bake), it is still, in a sense, working blind. Since the 1980s a slew of new instruments and manufacturing processes have emerged that enable scientists and engineers to see; that is, measure and manipulate at the nanolevel. It is because of these that nanoengineering is emerging as a commercial endeavor.

What follows is a review of the main ways that have emerged over the past 20 years for manufacturing nanomaterials and nanostructures of various kinds. In examining these approaches, the reader should note that some are top down and others are bottom up, two terms that are heard frequently in nanotech circles.

- In top-down manufacturing processes, products are designed using macrolevel materials. To put it perhaps a little crudely, one whittles away at the material until nano-level features can be achieved. While the whittling process is, of course, as old, even perhaps older, as mankind, this process does not inherently produce nanoscale structures. Its ability to do so depends on the material being used and especially on the tools being used. As we discuss, much of the buzz about the commercial nanotechnology today can be traced back to better tools that "sculpt" at the nanoscale. One area where this has become increasingly important is in the semiconductor industry where the big chip manufacturers are looking for better tools to handle ever smaller feature sizes on chips.
- In archetypal bottoms up approaches, products and materials are created one molecule at a time. Nanotechnology as originally envisioned was inherently bottoms up and this is the ultimate goal for most firms. Bottoms up creation is the way that nature apparently works and this is important, because much of the thinking in nanotech at present relies on biomimicry for its inspiration. However, nanotechnologists are practical people and must make do with the current generation of manufacturing technology, which is not especially bottoms up in nature. Also, manufacturing one molecule (or atom) at a time can sometimes be as painfully slow as it sounds, yet again another drag on the potential for real business opportunities.

Many of the techniques we discuss here come from the world of the semiconductor and electronics industries. This is because these industries are already working well within the realm of the nanoscale (under 100 nm) and have therefore pioneered many new ways of creating nanostructures. Traditionally, microelectronic circuitry has been created using optical lithography processes, but these are limited at the nano level by the wavelength of light. While the wavelength of visible and near visible light are tiny, if the features that it is supposed to create are tinier, the process ultimately becomes much like trying to create a beautiful sculpture with a chain saw. The obvious fix for this problem is to extend conventional optical lithography by using higher frequency light beams. For example, Intel has expressed a lot of confidence in what it is called extreme ultraviolet (EUV) lithography. However, there's only so far that you can go with this. The higher the frequency, the higher the energy. Trying to create a nanostructure with a high power beam can be a thankless task, since the energy may ultimately destroy the material. One alternative that has been suggested is to use electrons instead of photons. The result of this is e-beam lithography.

These approaches and others will be used to help the microelectronics industry create ever smaller structures. What follows is a brief description of some of the newer manufacturing techniques that are more than just an extension of conventional optical lithography and either are in commercial use today or are being seriously considered for commercial use. With the possible exception of self-assembly, most of the approaches listed here are intended primarily for circuit manufacture of various kinds. However, sectors outside the electronics and semiconductor industry that are increasingly looking at entirely new ways of creating nanostructures are also beginning to find a growing number of applications in areas that have nothing to do with electronics. Pharmaceutical applications are high on the list of where these newer nano-manufacturing approaches are already finding revenue-earning applications. There are also broader applications for these techniques if purely R&D applications are being considered.

It seems reasonable to believe that, as nanotechnology evolves, other parts of the nanotech sector will look to existing production technologies, originally designed for semiconductors, rather than reinvent the wheel.

The reader should note that the list that follows is not complete and that each of the approaches described below are frequently not a single technique, but rather a class of techniques that share some important characteristics. Nor are the approaches entirely mutually exclusive. Increasingly, multiple approaches to nanostructure creation will be used in the semiconductor business and it seems likely that this will also be that the way that manufacturing evolves in other sectors, too.

The list below is not exclusive and nanoproduction techniques are one of the biggest areas of new ideas and innovation in nanotech as a whole. So this list can be safely predicted to grow over time.

Scanning Probes The first tools to enable nanoscience (and hence nanotechnology) in a significant way were *scanning probes.* These emerged from work done by IBM's famous research laboratory in Zurich in the 1980s. This work was based on a simple phenomena that we all have experienced, namely, different materials exert different forces on objects that are dragged over them. Thus, if you run your hand over a polished table, it feels smooth, with just a little friction pulling your hand back. If you foolishly spill a bottle of syrup on that table and then run your hand over it again, it feels quite different. Your hand is dragged back by forces created by the syrup and which are quite different than the ones that existed before the syrup was spilled.

Imagine this basic idea transferred to the "nanorealm," and you have the scanning probe. Instead of your hand, there is a probe, or "tip," which itself is a nano device. As this probe is slid over a surface it measures the nanoscale phenomena based on its sensing of forces, similar to those described for the table example. However, while scanning probe instruments were originally thought of as measurement devices, they also possess some potential as manufacturing tools. Clearly when you run your hand over that syrup-soiled table, you are not only feeling the syrup deposit but changing its shape, perhaps by smearing it across the table. Similarly, the scanning probe tips can push atoms and molecules around or pick them up. One writer has described the scanning probe as "the earthmover at the nanoscale."[28]

Scanning probes have been used to manipulate objects at the nanoscale, mostly in the R&D environment and seem to have limited use for developing nano products in volume. Building nano products atom-by-atom or molecule-by-molecule would (almost literally) take forever. Scanning probes, of

Table 1.2
Scanning Probe Technologies

Type of Scanning Probe	Description	Comment
Atomic force microscopy (AFM)	Electronics is used to measure the force exerted on the probe as it is moved across a surface.	Collective term for a variety of measuring techniques used in nanotechnology and biotechnology.
Scanning tunneling microscopy (STM)	The amount of electrical current between the probe and the surface is measured.	Primarily for measuring either the local geometry or electrical characteristics at the nanoscale.
Magnetic force microscopy (MFM)	In this scanning probe technology, the probe is magnetic and is similar to the head on a disk drive.	Used for measuring local magnetic properties of nanoscale materials.

which there are several kinds (see Table 1.2) are also rather expensive. So other approaches to nanoscale manufacturing are likely to be a better bet. Nonetheless, scanning probes of some kind are likely to be found in many nano businesses, even those that are primarily aimed at developing intellectual property rather than making products for sale. This is because they are essential tools for nanoscale measuring in R&D and also for some kinds of prototyping.

While the potential for scanning probes in manufacturing is inherently limited, their ability to shift atoms and molecules is being enhanced with software that enable scientists and engineers to interact with the scanning probe through a 3D graphics workstation. In this setup, the user can actually see the atoms/molecules that he or she is moving around and with the addition of some kind of feedback mechanism may actually be able to feel them. In theory, the engineer or scientist using this approach could actually built nanostructures with their hands. But again, the process, fascinating as it is, is of virtually no use for volume production. Today, there are a score of scanning probe equipment manufacturers scattered throughout the world. The best known in the nanotechnology world is probably Veeco.

Dip-Pen Nanolithography Dip-pen nanolithography (DPN) is a system that uses one or more atomic force microscope probes to write on a surface much like an old-fashioned quill pen. Lines with widths below 10 nm can be drawn in a wide variety of inks. (The inks may be conducting so that electronic circuitry can be created with this process.) Key advantages of DPN include:

- High level of precision. Attaching a nanoparticle to a nanowire, for example, is quite achievable.
- No mask needed. Another major advantage of DPN is that it does not require a mask, thus eliminating one of the most expensive and problematic elements of other, more conventional lithographic processes. The masks used in semiconductor manufacturing can sometimes cost as much as one million dollars.

One problem with this approach to nanomanufacturing is that using it for volume manufacturing implies that very large arrays of tips must be deployed. NanoInk, a firm that has been a leader in dip-pen lithography, has demonstrated arrays of over a million probes, but these are not independently addressable, which limits the applicability of this approach. However, IBM's Millipede memory technology uses 4,096 addressable probes, which shows that quite a lot can be done with this kind of technology even at current the current state of technological development. Should it prove possible to develop very large arrays of independently addressable probes at a low enough cost then sufficient scalability might be found for volume manufacturing.

For the time being, DPN lithography seems likely to be used for (at best) moderate volume manufacturing or repairs. It is therefore favored for the manufacture of products that require nanoscale features, low production runs and rapid turnaround from design to production. These include:

- Rapid prototyping;
- Production of masks and masters;
- Research applications;
- Placement of nanotubes, nanowires, or other nanostructures;
- Precise "nanopatterning" of materials.

DPN has already seen commercial use for making repairs in the electronics for thin-film LCD displays and has demonstrated its potential for additive repair of photomask layers. The value of this tool for repair applications comes where products are of high value (e.g., photomasks) and/or have a high failure rate, as in the case of the thin-film transistors for large LCD displays. The increase in circuit failure rates grows as feature size diminishes, and the increase in masks costs suggest an increasing inherent value in activities such as repair and circuit editing.

DPN is also finding novel applications. One firm has found one of the first volume applications for DPN in providing encrypted manufacturing identification on pills. This is a method intended to guard against the increasingly common practice of drug counterfeiting. Also, as the nanomechanical machines begin to emerge as serious commercial product, DPN, may have some potential for placing glue or solder at joints of these and other complex nanostructures.

Nano-Imprint Lithography Nano-imprint lithography (NIL) covers a range of techniques for making nanoscale patterns and is largely interchangeable with the older, but less-used term, soft lithography. There are essentially two variants on nano-imprint lithography: inking, those making positive copies of an image, and imprinting or molding, those making negative copies.

Inking has much in common with the traditional printing press. A pattern is created by removing material to leave a bas-relief, which can then be inked and applied to a surface, where it leaves an ink duplicate of itself. The semiconductor industry experimented with this idea in the early 1970s but dropped it because of contamination and defects. These problems have now been ameliorated through the use of a softer material for the stamp, namely polydimethylsiloxane (PDMS). Inking approaches can produce features well below 50 nm and a wide variety of "inks" can be used.

Inking of this kind is already in use commercially in the creation of microfluidic systems (Surface Logix is a pioneer in this area). However, it seems that when NIL is discussed in the nanotechnology literature, it tends more frequently to refer to imprinting and molding techniques in which a harder stamp is used to leave a negative image of itself in a soft material into which the stamp is pressed or that is allowed to flow into the gaps around the image on the stamp. This approach offers better precision and alignment than inking and is already used commercially to produce subwavelength optics. One imprinting approach, "step-and-flash," is being vigorously pursued for electronics by the Molecular Imprints with the active support of Motorola, which is both an investor in the company and the first customer for its step-and-flash equipment. Another nanoimprint variant is laser-assisted direct imprint. This uses a patterned quartz master directly on silicon and an excimer laser to briefly melt the silicon, creating in one step and less than a millionth of a second patterns that normally require multiple steps and many minutes, and with feature sizes down to 10 nm.

In addition to Molecular Imprints, another company already selling nanoimprint tools aimed at electronic circuit manufacture is Nanonex, whose nanoimprint technologies have been commercialized for some time in the production of optical components by NanoOpto. Other companies have been making nanoimprint tools for many years. Obducat of Sweden is the best-established company devoted to nanoimprinting and it has found applications for this technology in the semiconductor, sensor, data storage and electronics markets. Advocates of the imprinting and molding variety of NIL note that it is different from other printing processes in that it is "primarily a physical deformation process," and does not use an ink so that it "avoids many problems in other lithographical methods."[29] More generally, the characteristics of NIL that make it so attractive for nano creation are:

- Features created with NIL can be truly nanoscale. They can be reliably created down to below 20 nm and perhaps to under 10 nm.
- Good economics. The low cost of creating masters makes small production runs economic, but reasonably high throughputs are also possible.

As with all nanoscale production methods available to day, NIL has problems that have yet to be resolved. Defects are one such problem that has kept NIL from becoming a serious contender for high-volume applications.

Printing One of the biggest opportunities that is likely to emerge in nanotechnology (understood broadly) is in so-called printable electronics. This technology, which we will discuss more fully in Chapter 3, could produce some genuinely novel products, such as large computer displays that can be rolled up like paper, RFID tags that can be printed on packaging like bar codes, and solar panels that can be laminated directly on walls. The basic idea here is to use a printing technology to create low-cost electrical circuitry on a plastic, glass, paper, or other substrate of some kind. (Hence printable electronics is also sometimes called plastic electronics.)

The printing technology used to create this kind of circuitry can be of a number of kinds, including traditional offset, gravure, and flexographic methods. It could be NIL, for example. Indeed, if genuinely nanoscale features are required, this may be the only way to go. Ink-jet printing is also a popular option, since it does not need a mask and so can print electronics products in very small quantities.

Printable electronics in its current version does not create at the nanoscale in the truest sense. However, the inks that are used for printable electronics are often nano-engineered. In particular, nanometallic silver inks are already being used to print antennas for RFIDs and will probably also be used to print the backplanes for flat-panel displays. The other kind of ink being tested out for printable electronics is based on organic molecules, most especially polymers. To date these are not nanoengineered, but rather are built using the processes of conventional plastics chemists. However, new applications are likely to demand novel inks and it seems quite likely that future organic inks will be created using nanotechnology and even current ink engineering involves designing molecules, which certainly has a nanotech flavor to it.

Is printable electronics really nanotechnology, since the feature sizes of printable electronics are well above those that would usually qualify as nanotech. Taking the pragmatic approach that we hold throughout this entire book, we are including printable electronics under nanotechnology on the grounds that it is about novel manufacturing materials based around new advanced materials designed, at least to some extent, at the molecular level. Also, it can be claimed that a key objective of this form of manufacturing is to move it closer to the "nanocosm." Indeed, one of the problems with using ink-jet printers is that line widths from standard industrial printers are typically too wide to create complex circuitry. Consequently, at least one firm, Litrex, is developing special ink-jet printers that are more suitable to finely created circuitry. In addition, feature sizes can be improved by combining ink-jet with other technologies. Plastic Logic has used an approach combining ink-jet printing and repellant forces in materials to get feature sizes down to 100 nm.

Ink-jet seems to promise a radically new direction for manufacturing, and ultimately the basic approach could be extended well beyond the manufacturing of two-dimensional circuitry. Some futurists suggest that in the future tiny fabs based on this technology and capable of printing in three dimensions will be able to manufacture products according to customized and personalized blueprints. A little fantasy story that is sometimes told is of a future home where a remote control for a television is lost, but is replaced with one built using the family ink-jet fab. Don't count on any of this happening any time soon. The immediate impact is likely to be a lot less dramatic. In the meantime, printable electronics provides a welcome diversion from the cruel logic embodied in today's semiconductor industry in which (as everyone knows) processors and logic get ever more powerful, but in which (as only relatively few people know) leading edge manufacturing plants are escalating in cost to a point where they will soon cost the equivalent of the GDP of a small nation.

Self-Assembly and Molecular Manufacturing Self-assembly is the production process that most typifies nanotechnology and is radically different from all the other production techniques covered in this chapter. It is suggestive of the Drexlerian notions of nanotech and these certainly qualify as self-assembly. However, self-assembly does not necessarily imply anything as futuristic as Drexler's molecular assemblers. Growing crystals, which is something that every high school chemistry student has done, is also self-assembly.

At its core self-assembly implies designing structures, at the molecular level and up, using materials that will spontaneously assemble themselves into the desired structures. The paradigm here is obviously a biological one: the development and growth of animals and plants being the most impressive example of self-assembly that anyone can imagine. However, the reason for adopting self-assembly in nano-manufacturing is not biomimicry for its own sake:

- *Abundance.* The biggest attraction of self-assembly is that it can lead to unlimited parallelism in production. If you can find a suitable material that can serve as a feedstock, all you have to do to scale up production by ten times is start with ten times the amount of feedstock.
- *Uniquely Suited to Nanotechnology.* While all the other production technologies reviewed here are well-suited to nanotechnology, only self-assembly is completed unlimited when it comes to operating at the nanoscale. With the other techniques the question as to just how small features can be made is always raised. With self-assembly, the nanocosm

> is the easiest place to start. Molecules naturally create nanoscale structures.

This is not to say that self-assembly is without problems. Perhaps the biggest issue is making the connection between the nanostructures built using self-assembly and the everyday macroscopic world. For example, it may turn out to be relatively easy to produce complex electronics components using self-assembly approaches, but adding electrical contacts and placing nanoscale components in a system that otherwise uses conventional components can be a problem. What is needed is a way to integrate components with features at the level of a few nanometers or less with structures with smallest features a hundred times larger. Accurate placement can be achieved with the use of scanning probe tools but there is little prospect that this way of doing things could be scaled up to mass production.

Another issue with self-assembled structures is that they tend to have a relatively high fault rate. This isn't intrinsic, just hard to avoid. As a result, many research groups working on self-assembly projects are developing fault tolerant approaches in their work. It should also be noted that while biomimicry is an important inspiration for nanotechnologists focusing on self-assembly, natural systems tend to radically different in design and operation from typical human engineering and we still have much to learn about how these systems work.

The business opportunities that stem from self-assembly depend a lot on how you view self-assembly. If it includes crystallization processes, then the market for self-assembled products already runs into the billions of dollars. Similarly, vapor deposition processes that are used to create optical fiber, some organic light emitting diodes (LEDs), and a variety of other products may also be thought of as self-assembly and again already generate huge revenues. However, if one thinks of self-assembly in Drexlerian terms, that is, in terms of self-assembly of complex systems, there won't much commercial activity for many years (or decades) to come.

There is some promising work being done at universities on building blocks that may ultimately lead to molecular assemblers and also some indications of technology directions that early complex self-assembly might take. For example, the version of self-assembly that seems to offer the most promise is "templated," or "directed," self-assembly. The basic idea here is to make devices with an area appropriately structured to guide self-assembly. An open question is the degree to which organic molecules will serve a useful role in self-assembly. As we have already noted, biomimicry may be more inspirational than practical in nanotech. The most versatile chemicals for self-assembly are biomolecules such as proteins and especially DNA, but these molecules are generally easily damaged and researchers often use them as a template for laying down, for example, metals, carbon nanotubes, nanowires or quantum dots, rather than as the self-assembling molecule itself. Meanwhile, the closest commercial processes we have today to advanced molecular manufacturing stop a little short of self-assembly. And Zyvex, a successful nanomaterials firm that has a well-publicized molecular assembler projects stresses that it is a "very long-term project," which is surely no less than the truth.

Summary: Key Takeaways from This Chapter

To summarize then, nanotech is about designing at the molecular or atomic level. Put this way, nanotech is a lot less exciting than the visions of rampant molecular assemblers that one reads of in Drexler's books or for that matter in Michael Crichton's *Prey.* It doesn't even seem all that new. As cynics are quick to point out, nanoproducts such as carbon black have been in use for a long time. The semiconductor industry has in a sense between engineering at the quantum level for more than 30 years now.

All this is undeniably true, but misses the point. The reason why nanotechnology is considered by so many as the next big business opportunity is not because it is an entirely new idea, but rather because the ability to nanoengineer products and materials has now reached a level of sophistication that it can have a broad impact in many sectors of the economy. And the reasons for this development are the range of new tools and manufacturing processes that we have discussed in this chapter. The main things to remember from this chapter are:

1. A useful working definition of nanotechnology is the one used by the U.S. National Nanotechnology Initiative, which specifies nanotechnology as occurring on the "atomic, molecular or macromolecular levels, in the length scale of approximately 1 to 100 nanometer range [and] creating and using structures, devices and systems that have novel properties and functions because of their small and/or intermediate size."
2. The smallness of the nanoscale is important because at the 1 to 100 nm scale quantum mechanical and statistical mechanical effects have an impact that they do not have at the macro level. These effects may present a new opportunity for nanoengineers, as is the case with

quantum computing, or a problem to be solved, as is the case with soft errors in computer memories.

3. Nanotechnology has a history going back to the theoretical musings of the physicist, Richard Feynman in the late 1950s and discussions of molecular manufacturing by Eric Drexler in the 1980s. But it has come to mean a new kind of engineering occurring at the molecular of atomic level. This kind of engineering has taken on growing commercial importance as new tools and manufacturing processes have come into being enabling practical nanotechnology. These include scanning probes, dip-pen lithography, nanolithography, and self-assembly. It is really the availability of such tools that has made widespread commercialization of nano-enabled products possible for the first time.

Further Reading

The following books and articles are intended to provide more detail on the topics raised in this chapter. They are not intended as a complete bibliography of the areas covered, but just a way for the reader to take any issues that particularly interest him or her to the next stage. The books marked with (T) are more technically oriented.

Basic Books on the Technology of Nanotechnology There are already a growing number of books that purport to explain the technology in nanotechnology. There will probably be more by the time you are reading these words. According to Amazon.com, the most popular of these books is *Nanotechnology: A Gentle Introduction to the Next Big Idea*, by a father-and-son team, Mark and Daniel Ratner.[30] This is a quick read—one could probably get through it in a couple of evenings. It is well written and does an excellent job at conveying the kind of thing that nanotechnologists do. It has some things to say about commercialization and business issues too, but not much. It has received some criticism for focusing too much on the work done at Northwestern University where one of the Ratners is a professor. The best survey of the technology for the serious reader, but one which does not require more than a good high school science education is *Nanotechnology: Basic Science and Emerging Technologies.*[31] This was written by a team of Australian writers and is both broad in its coverage of matters nanotech and deep in dealing with science behind them.

Books and Web Sites on the History and Current State of Nanotechnology. The two classics of nanotechnology are Feynman's "There's Plenty of Room at the

Bottom" lecture, which can be found in many places on the Web but is most accessible at http://www.its.caltech.edu/~feynman/ plenty.html, and *Drexler's Engines of Creation: Challenges and Choices of the Last Technological Revolution.*[32] It is fair to say though that these writings are now only of historical interest. The transformation from the Drexlerian vision of nanotechnology to the business-oriented vision is described in a number of books and articles. But perhaps the best survey of what happened is "The Incredible Shrinking Man," an article that appeared in *Wired* magazine some time back.[33] For a current perspective on nanotechnology from a Drexlerian perspective, see A. Storrs Hall's *Nanofuture: What's Next for Nanotechnology*,[34] For an aggressively anti-Drexlerian view, see *Nanocosm: Nanotechnology and the Big Changes Coming from the Inconceivably Small.*[35] Although a little out of date now, this book gives a very good feel for mainstream nanotech R&D work. One other book that is widely available in bookstores that purports to be an overview of nanotech commercialization is Jack Uldrich and Deb Newberry's *The Next Big Thing Is Very Small: How Nanotechnology Will Change the Future of Your Business.*[36] This is a good popular survey, but is more of a cheerleading exercise than a serious analysis of market opportunities. Read it to boost your spirits, not to get business ideas.

Finally, while there are now numerous Web sites dedicated to nanotechnology. One that is definitely worth a look is http://www.nanotechnology.com. The pioneer of nanotechnology Web sites is http://www.smalltimes.com, which also has an associated magazine.

More on Nanotools Most of the basic books on nanotechnology, discuss nanotools only in passing, although the Ratners' book mentioned provides a pretty good overview. The definitive work on nanotools has yet to be written, but until it is, *Alternative Lithography: Unleashing the Potentials of Nanotechnology*,[37] which is edited by Clivia Sotomayor Torres, is a good up-to-date survey of at least some of the field.

2

The Business of Nanotech

> The ultimate nano-product is a bionanoparticle-infused nano-engineered nutraceutical that is delicious, non-fattening, makes you thin and look great [and] keeps you young and healthy
>
> —*Darrell Brookstein in his book* Nanotech Fortunes[38]

Introduction: The Hidden Assumption of Nanobusiness

As someone who tracks the commercialization of nanotechnology, I spend much time talking with the engineers, product planners, and business development managers whose job it is to bring nanotechnology into the real world. The impression I am left with is that there are many firms who have now nearing the end of the difficult task of completing a stable materials/technology platform with interesting properties, but who have little idea of where to take this platform next. Commercialization is the next step and they can see myriads of applications for what they have developed, but they simply don't have the marketing, production or financial resources to take their new platform and commercialize it for several highly distinct sectors.

A new product that can potentially generate revenues in (say) the computer display, drug delivery, natural gas, agriculture, *and* aviation industries is an exciting product. It can also be a frustrating product, because opportunities and strategies vary from one sector to another. Making money with nanotech in the energy sector is different to making money in the semiconductor sector. There are, however, some common business issues that seem to pop up in all sectors of nanotechnology. To return to a theme in the previous chapter, this is one of the factors that gives unity to the nanotechnology as a business. My own market research and strategy consulting work and that of my firm, NanoMarkets LC, has ranged over the electronics, semiconductors, energy and pharmaceutical industry. What follows in this chapter is a distillation of the wisdom I have discovered from talking with scores of executives in the budding nanotech industry.

Industry is a word that those in the business of nanotechnology don't like to use that much. A *BusinessWeek* cover story devoted to a survey of nanotechnology makes several references to a nanotechnology industry, but then goes on to note, if somewhat paradoxically, that "Nano is not a single industry but a scale of engineering involving matter between 1 and 100 nanometers." At some level it is obviously true that nanotech is not an industry in the way that the automobile industry is an industry. The automobile industry turns out automobiles. Even if we could discern something that could reasonably called a nanotech industry, it would not be churning out "nanotechs," as it were. Noting this, conventional wisdom has it that nanotechnology is an enabling technology that affects (or will affect) numerous firms in many industries. Firms that use nanotech in their processes or products include textile firms, energy firms, semiconductor firms, automobile firms, and so on. Taken together they still do not constitute a nanotech industry.

In deference to this consensus point of view I have tried to use the term "nanotechnology sector" in this book, rather than "nanotechnology industry." This seems nice and neutral! Of course, in many ways this is just semantics, hiding the more interesting question of why it is reasonable to think of commercial nanotech as one thing rather than many. That this is the case is evident that we have a burgeoning number of commercial nanotech conferences, organizations, consultancies, and books, such as this one.

Where does this unity come from? While it may be self-evident that the nanotech sector is not an industry like the automobile industry, the claim that nanotechnology is an enabling technology applying to a wide variety of industrial sectors, doesn't really help that much either. This is because, examined carefully, nanotechnology actually seems to be a collection of many different technologies. Thus, for example, although they both fall under the definition of nanotechnology, the application of spintronics (see next chapter) to making memory chips has little connection—at least from a commercial perspective—to the application of carbon nanotubes to making stronger frames for aircraft, for example.

So what we seem to be left with then is that the nanotech sector consists in a variety of commercial activity in many different areas using many different technologies, with these technologies sharing nothing more than the fact that they operate at the level of the very small. More explicitly, there seems to be a hidden assumption of nanobusiness that firms who are skilled in nanoengineering will be able to design and ultimately sell products into sectors that traditionally have had little to do with each other, based on a common materials/technology platform. It is understood that the marketing skills in (say) the energy sector will be different from those needed in the textile sector, but the hidden assumption of nanobusiness essentially claims that this is a secondary matter compared with the value implicit in a nanotechnology platform to which a firm has a proprietary right and which functions as a broad enabler in many different areas. This is the philosophical underpinning of one of the common business models found in the nanotechnology sector, that of the "pure IP" model, which, as we shall see later in this chapter, is a fairly common business model for nanotechnology. In this model, a firm develops a platform with interesting characteristics and then sells off rights to use this platform to firms with marketing and manufacturing skills appropriate to a specific economic area.

It's hard to quarrel in general terms with the hidden assumption. It is reinforced by the view that we presented in the previous chapter that, no matter whether we are talking about nanoelectronic circuits, new kinds of nanocomposite materials, nanobiological devices, or one of the many other kinds of nano-engineered product, the nanoengineering is likely to be carried out with a fairly narrow set of tools. However, while true, this kind of description can be a little unsatisfying if what you are looking for is a some idea of what the business characteristics and structure of the nanotech industry will look like over time.

Three Scenarios for the Nanotechnology Industry

This is not a semantic issue, nor an academic one. If nanobusiness really is a lot of different things posing under one hat, with little to unify it except a faddish nanotech moniker, then the nanotech sector will dissolve in a few years and perhaps even the term "nanotech" will disappear from the vocabulary to be replaced by some other yet-to-be-invented terminology. As I will suggest below, this is actually a possible scenario and should be taken into consideration by any firm that wants to brand itself as a nanotech firm. In what follows, I want to consider this scenario, along with two others for the future of the nanotech sector, along with the likelihood that one might reasonably assign to each scenario and how each might impact the way business is done by nano-oriented firms.

I have spent a good deal of time thinking about why nanotech will remain a unified sector. I don't have a perfect answer. In fact, as I have already noted above, I think there is a finite possibility that it could turn out to have been something of a fad. But I think there are also some hints of why nanotech will stay nanotech, as it were.

Scenario # 1: Nothing Special About Nanotech One answer, which may seem initially appealing, says something to the effect that nanotech may not be like the automobile industry, but it *is* like certain other industries that cater to a very broad range of end user segments with an equally broad range of products. To the extent that such industries exist, one could argue, there is nothing special about the nanotech industry.

In looking for existing industries from which one may draw such an analogy, one might select the chemical industry, which sells everything from bulk sulfur by the ton to thimbles of specialty chemicals. It sells these chemicals to end users ranging from the building industry to the pharmaceutical industry. Another similar industry is the semiconductor industry, where it is also typical for firms to use similar technologies but produce different products and sell into different spaces, which again sounds a little like the way that people describe commercial nanotech. Texas Instruments and Intel are, for example, both huge semiconductor firms using similar manufacturing approaches to producing microcircuitry, but they manufacture very different kinds of circuitry and have different kinds of customers. Intel's core business is selling chips for personal computers. TI's business lies in the area of digital signal processors for audio and video equipment and in making a huge variety of analog chips.

Such analogies might be thought of as existence proofs that something like the nanotechnology sector could remain stable over time and emerge as an industry in some sense. Unfortunately such analogies can be taken only so far. For example, push the analogy between the semiconductor sector and the nanotech sector too far, and it quickly unravels. Despite their great and obvious diversity, semiconductor products are all quite similar to each other: they are all a class of computer component for all intents and purposes. In addition, while it is true that these semiconductor products may be sold into many different markets, in any given market they compete with very similar semiconductor products from rival firms. A similar analysis could be performed comparing nanotech with the specialty chemicals sector with similar results.

Now compare the nanotech sector. The products, or at least the potential products, of nanotech are diverse in an entirely different way from those of the semiconductor industry. They cover devices, materials, and many other kinds of structures and while similar nano products from different firms may compete with each other, many of today's nanotech firms are supplying new nano-enabled products into spaces that are well served by a range of entirely different non-nano products. Nano-enabled cancer treatment products compete in the marketplace with non-nano cancer treatment products. Indeed, most end users (patients and doctors) will not give a hoot that the product is nano-enabled.

Once again, what all this tends to indicate is that today's nanotech sector is probably more diverse than any other industrial sector that we can imagine. This is more evidence, that perhaps, given half a chance, the nanotechnology segment will explode into lots of little fragments and will never be a self-respecting industry.

On the other hand, sometimes the world doesn't quite work out the way that rational analysis suggests it should and it is perfectly possible that the analogy with the chemical industry or the semiconductor industry may turn out to be strong enough to bring about a permanent, identifiable, and stable nanotech sector. For this to happen there would have to be some binding force that keeps the industry together. This could be its ties to government nanotech programs (see below) or perhaps some sense of community built around a major organization or trade show. As with all the scenarios we explore here, the "nothing special about nanotech," scenario has profound implications for industry structure and how money is made in nanotech. As far as structure goes, the sector would consist of firms that would identifiably be nanotech firms. These might be entirely new firms that have specifically been established to compete as nanotech firms or they may be older materials, chemical, or other firms who have transformed themselves into nanotech firms in order to keep up with the pace of technological change. As far as the business models for the nanotech sector under this scenario, they must be able to support a very broad range of activities in the line with the very broad range of applicability of nanotech itself. This may favor a model in which IP licensing may be dominant (perhaps with some extra revenues derived from technology transfer fees). This is really the only way that most nanotech firms will be able to compete in all the sectors that are required of them under this scenario. For the very largest firms, such as the multinationals that rebuild themselves in the nanotech image, a combination of IP licensing and competing in numerous applications sectors may also be possible.

What is the likelihood that this scenario could actually turn out to be the way that the nanotech sector develops? My guess would be that it is fairly unlikely. This is for two reasons. The first is that the centrality of the IP model to this scenario may make it unsustainable. I will have more to say about this later in the chapter, but the basic point is that the very versatility of nanotech will, I believe, make IP less of a barrier to entry in nanotech than some people, most especially, IP lawyers, currently think it will. In this sense, this scenario for the nanotech sector contains the seeds of its own destruction. The other reason for thinking that this is not the direction that the nanotech sector will ultimately take is that it is hard to imagine what the institutional factor that will bind the nanotech sector together would be. Many (but not all) of the trade shows and organizations that have been established to cater to the needs of nanobusiness could not be said to be thriving and the government programs seem targeted to promoting nanotech in some specific sectors, but not really to building a nanotech industry per se.

Scenario # 2: Things Fall Apart As I have already hinted, there is a possibility that nanotech may turn out to be something of a fad. By way of example, as to what I have in mind, let me cite "multimedia networking," a sector that I used to follow as an analyst and newsletter publisher, about a decade ago.

The term "multimedia networking," is seldom, if ever, used these days, but this sector was supposed to combine content from the entertainment and television industries over networks built with technology taken from cable television, computer, and telephone industries. The forces that were supposed to be creating this industry were deregulation of the broadcasting, cable TV and telephone industries, along with the increasing digitalization of the infrastructure in all three of these industries. These factors were supposed to lead to the convergence of computing, television, and telephony. Eventually, it was said that all the firms in these sectors would use similar equipment, adopt similar technical standards, and (most importantly) get into each other's business.

But today, nobody talks about multimedia networking.. The big conferences and trade shows that focused on multimedia networking are gone and so are the standards organizations. Cable television and telephony use different technologies and types of equipment, just as they always have. Sometimes this is the case even when they are supplying more or less the same services to customers, the best example being broadband. For my own part, the newsletter that my firm published on multimedia networking flourished for a couple of years and then fizzled out.

Now, many of the services and products that analysts expected to come out of the multimedia networking industry in the early 1990s *did* actually come into being. Firms and individuals made, and continue to make, money from their success. But the supposed unity of the sector just wasn't there. The thinking and interests of cable companies, telephone companies, and computer companies were just too different to create an entirely new sector. Firms that went into multimedia networking believing that they could sell software and equipment to (say) the cable and telephone industry were generally unsuccessful. Cable companies stayed wedded to their technology platforms and telephone companies to theirs. When the telephone companies decided to go into the television content business, it was a disaster.

It is all too easy to imagine nanotech going down the same path, although this book is written in both the hope and assumption that this is not going to be the case. Suppose, however, that I am wrong and nanotech follows the same path as multimedia networking. Then ten years from now, many of the market opportunities that I describe in these pages will have materialized. It is possible that (say) molecular memory, new materials for photovoltaics and new kinds of drug delivery systems operating at the molecular level will all be generating lots of revenues. No one will think of them as having much to do with each other. No one will think of a firm producing a nanoengineered drug delivery system as having much in common with one producing a nanoengineered catalyst to improve the efficiency of a hydrocarbon fuel. And no one will remember the word nanotechnology, except perhaps as a fond memory.

This may seem unimportant—no more than the loss of a word. But if the "things fall apart" scenario is the way things pan out, one of the implicit claims that are currently being made for nanobusiness will be proven false. The progress in physics, biology, and material science that have made nanotech a possibility will be as a real as ever and many of the commercial products that are now being, or soon will be, produced by firms under the nanotech moniker may also succeed. Under the "things fall apart" scenario, there will be no nanotechnology sector as such and no uniquely nanotech-flavored strategies or business models that enable businesses to bridge many very different traditional sectors with their product and marketing strategies, based on a common proprietary materials/technology platform.

What is the likelihood that this is the scenario that will pan out? The probabilities must be judged as moderate. If you are reading this book, the chances are that you already have something of a belief in the future of nanotech. Nanotech is still too new to judge it a sure bet. There again, you may be reading this book in a secondhand bookstore years after it is written and smiling to yourself about how anyone could be so silly as to believe that something like nanotech could ever constitute a industry.

Three Sectors to Rule Them All The first scenario described above has nanotech as a unified sector. The second scenario that we have discussed portrays nanotech as no sector at all. The logical third alternative technology is that the future of nanotech may be in the form of a relatively small number of sectors.

This is more or less the view that I am going to take throughout this book. More specifically, I am going to take the position that the vast majority of what is today being characterized as nanotech really falls into three areas: nanoelectronics, nanobiotechnology, and nanoenergy. Nanoelectronics encompasses both electronics and semiconductor industries. What I am calling nanobiotechnology includes medicine, healthcare, pharmaceuticals, and life sciences more generally. The nanoenergy sector that I have in mind covers fossil fuels, alternative energy sources, and energy sources for mobile electronics.

These three sectors overlap in important ways and are, in the strictest sense, arbitrary. They all use the tools as described in Chapter 1, but each has some fairly well defined business characteristics associated with it, which I will describe in later chapters. That the nanotech business can be easily broken into three sectors may be hard to swallow for those readers who have read some of the more breathless accounts of nanotechnology, such as Jack Uldrich and Deb Newberry's book, *The Next Big Thing,*[39] which portrays nanotech as impacting just about every segment of the economy.

In fact, this view of the future ultrapervasiveness of nanotech is largely correct in my opinion. However, the impact is likely to be felt from developments that can roughly be shoved into the three "uber-categories" I have described. For example:

- As we discuss in a later chapter, agriculture can be expected to be a prime beneficiary of nanotech but mainly through developments in nanosensors, which are nanoelectronic devices.
- The automotive sector is expected to be a major beneficiary of nanotechnological developments, but many of those developments will be directly related to new and/or better power sources
- The economy in general will benefit from improvements in health due to nanomedicine.

Accounts that emphasize the extreme diversity of nanotechnology may therefore miss the point. They are not actually mistaken—complex nanotechnology products certainly *do* have the potential to affect every level of the economy from mining to banking services, but primarily through the medium of its impact as an enabling technology in electronics, life sciences, and energy.

I have already laid my cards on the table and said that, of the three scenarios presented here, this is the scenario that I see as most likely for the future of nanotechnology. I do not, of course, have absolute proof of this. However, I do have some anecdotal evidence. First, the market research and consulting work that my firm NanoMarkets has done over the past couple of years suggests that at least 80 percent of the likely impact of nanotechnology will be found in the big three sectors of the economy. Second, in conversations with well-placed nanotechnology executives and journalists covering nanotechnology, I have gained the distinct impression that nanoelectronics, nanoenergy, and nanobiotechnology are going their own way in terms of media coverage and the conferences and trade shows that cover them. In some ways, they are becoming specialized subsectors or the electronics/semiconductor, energy, and life sciences sector.

Now, one is apparently supposed to feel a kind of awe at the many different implications of nanotechnology, but if the "three sectors to rule them all" scenario pans out, a cynic might be forgiven if his or her reaction was almost the exact opposite. After all, if nanobusiness splits into three sectors, a critic might ask how can nanobusiness really retain a structure as a unified sector? Such a reaction may well turn out to be prescient. However, the view that will be taken in this book is that three subsectors is sufficiently few to avoid the whole notion of nanobusiness dissolving and that there will be sufficient overlap in terms of technology, tools, and even products that nanotechnologists and business people in one subsector will be interested in talking to nanotechnologists in another subsector and even in moving from one subsector to another during the course of their careers. In evolutionary terms, we may think of the future of nanotechnology under this scenario as one in which it splits into races, but not into separate species.

Although there will therefore be important linkages between all the branches of nanobusiness, the three big sectors will vary in quite significant ways when it comes to business model. This is because each of these segments will be driven by the necessities of the big forces that we discussed briefly in the preface.

In this nanoelectronics sector, the most important of the forces that impact business strategy is mobility. I do not know what proportion of electronics activities is geared towards mobile communications and computing these days, but it is quite high and some of the largest electronics firms are now telling us that the next big thing is electronics is pervasive computing. The opportunities for nanotechnology will be shaped by this trend in the electronics, meaning that they will lie in areas such as nonvolatile memories, flexible displays, and other such products. The computer revolution was, of course, enabled by the development of the microprocessor. But what truly gave it momentum was that the products of the revolution—word processors, minicomputers, and PCs—fitted in beautifully with the spirit of those times in which people were looking for more leisure and more interesting jobs. Microprocessors did not create the information age. Instead, they enabled people to do what they wanted to do. If nanotech is ever to fulfill the dreams of its biggest advocates it is going to have to tap into the same kinds of societal megatrends that the computer industry did.

We should note that the need for nanotechnology in the electronics sector is also derived from the internal battle that the industry is now fighting to make sure that Moore's Law can be pushed forward to smaller feature sets and devices. We will have much more to say about this in Chapter 3.

In the life sciences area, the major megatrend is the aging populations in the developed countries who are, in effect, looking for the fountain of youth. This trend is now being propelled as a very large population group: the baby boomers. Today, the first baby boomers are about to turn 60. There can be little doubt that because their generation is so large, the needs and eccentricities of the baby boomer generation have shaped consumer markets since the end of the Second World War. The boomers have always been demanding and have tended to see themselves as somewhat beyond the reach of the rules that have governed previous generations. It's a safe bet that as their mortality catches up with them, the boomers will be a fast-growing market for both regenerative medicine and quick fixes for diseases.

The fountain of youth is a dream that is thousands, if not tens of thousands, of years old and will be with us long after the last baby boomer has shuffled off this mortal coil.This presumes, of course, that the last baby boomers don't turn out to be the first immortals.[40] Nanotech has a huge role to play by providing new materials for speeding and improving the quality of healing and making better body replacement parts. Novel drug discovery and delivery systems are also a prime area for nanoengineered products, especially since the days of the blockbuster drug, on which the pharmaceutical industry has relied for so long, appear to be over. Yet NanoMarkets' research indicates that the pharmaceutical industry, and perhaps, healthcare in general, are still very skeptical about the capabilities of nanoengineering. We expect this to change as the huge potential size of the addressable markets become apparent.

What nanotechnology can do to extend and improve life will be reviewed in Chapter 5, but there is little doubt that the megatrends indicated above will shape the life science part of nanobusiness for the foreseeable future.

Finally, there is the contribution of nanotechnology to the energy sector to be considered. I examine this in detail in Chapter 4. The high cost of energy is a problem for *all* industry sectors and it is clear that any firm that can provide an assist in this regard can make a lot of money. Nanotechnology seems to have much to offer here. Nanocatalysts can make hydrocarbon fuels last longer and nanoengineered products can help breath new life into the alternative energy sources such as solar and wind power. Further in the future lies vastly nano-enabled improved electricity storage and transport. For example, engineers have envisioned very low-loss electricity transmission over lines built with carbon nanotubes. Using nanotechnology to make the generation, storage, and transmission of energy better is a very diverse field, encompassing everything from nanosensors to monitor fuel cells and power stations to nanoengineered filters of various kinds. However, with demand for oil increasing as India and China's economies rise and with an unstable political situation in the Middle East, it is easy to believe that there will be considerable financial support for viable nano-engineered energy solutions, both from governments and from private industry.

The most newsworthy aspects of nanotechnology in the energy sector are likely to be in the improvement of electricity generation, storage, and transmission and in better engines for transportation of various kinds. Nanotechnology is also likely to have considerable impact on improving the efficiency of alternative power generation sources, most especially photovoltaics. However, a critical market for nanoengineered power sources including photovoltaics, fuel cells, and batteries, is also emerging in the mobile computing communications and computing area, where the absence of a suitable power source to drive multifunctional systems is now a serious drain on the market.

Thus the nanoenergy sector is being driven by two of the three megatrends that I identified at the beginning of this book, that of the energy crisis and the trend towards enhanced mobile electronics. In closing this section, I should mention that I did consider including a fourth trend as being all important in shaping nanotechnology opportunities, that trend being the growing need for security. Perhaps this is the appropriate place to briefly discuss this trend and why I am not giving it the prominence that I am giving to energy, mobility, and life sciences issues.

Ever since the 9/11 attacks, homeland security has become a dominant theme in world politics. This has only served to enhance a concern for data security that has been growing for years and is now one of the hottest topics in data communications. Anyone who has had to cope with a virus, Trojan horse or even adware on their computer will not need much of an explanation of why data security is such an issue and, of course, a heightened interest in data security is also a direct result of greater homeland security and military concerns.

Nanotechnology can make important contributions to improved security. NanoMarkets' research indicates that the most commercially important way that it is likely to do so is through the use of nanosensors, which can be (potentially) more sensitive, more ubiquitous and less costly than other types of sensor. In addition, nano-enabled quantum encryption is already bringing almost unbreakable encryption to the defense, aerospace, and financial services industries, and once its price comes down, will be much more widely deployed.

And yet, and yet. It is unclear that the current concern with security is something that will persist long term. I have talked with aerospace executives who have told me, off the record, that they do not expect the security trend to be a long-lived one, unless the U.S. suffers another major terrorist attack. While network and data security software is in constant need of updating, there does not seem to be any reason why there should be a great surge in network and computer insecurity in the immediate future, and, therefore, no trend compared to those we have identified in the electronics, energy, and life sciences areas.

Nanomaterials and Nanodevices

One of the virtues of the "three sectors to rule them all," scenario is that it goes some way to pinning down what nanotechnology *is* by stuffing it into a few reasonably well-defined categories. Another approach with a similar kind of reasoning behind it is to categorize nanoproducts into nano*materials,* nano*devices,* and nano*systems*:

- *Nanomaterials* are the simplest type of nanoproduct. They include nanopowders, nanotubes, nanowires, nanocoatings, and so on. They are typically seen as the low-hanging fruit of nanotechnology. They are the first nanoproducts to hit the market and they are probably the area where the first nanotech fortunes are going to be made. However, like all pure materials plays, nanomaterials will ultimately become commoditized and money will be made through economies of scale, just like it is today in the specialty chemicals and materials sector. Examples of nanomaterials include carbon nanotubes and nanometallic inks used for printing certain kinds of electronic circuitry.
- *Nanodevices* or nanostructures are more complex products, such as sensors or memory devices, built with nanotools and are likely to be using nanomaterials to some degree.
- *Nanosystems* or nano-enabled systems are macro-level products that are enabled in some significant way by nanomaterials or nanodevices in some significant way. This may be a pair of spill-resistant pants or it may be a new diagnostic system that employs a specially sensitive nanomaterial. It is probably stretching a point to include a missile just because it includes an MRAM chip somewhere or a car because it uses nanoengineered specialty glass for the windshield.

I will have a lot more to say about nanodevices and nanosystems than I will about nanomaterials. This is because it is in these two areas that I see most of the medium-to-long-term business opportunities being found and they are also the areas where it will be possible to attain the most sustainable competitive advantages, as is usually the case with more complex products. (There are more ways to create value added and barriers to entry in areas where the products are complex than in areas where they are relatively undifferentiated, as is the case with materials.)

Nonetheless, as I have already noted, the earliest opportunities in nanotechnology that generate sizeable revenues are likely to come from the nanomaterials segment. Establishing yourself as *the* place to go for carbon nanotubes is an opportunity that is here and now, but once the major suppliers of nanotubes in commercial quantities are in place (and that will happen soon)—large barriers to entry will be start to emerge and the window of opportunity will have past.

The early opportunities in nanomaterials are there because many other parts of nanobusiness, those higher up the value chain, will not be able to function unless there are good sources of nanomaterials. For example, carbon nanotube coatings and (to a lesser extent) carbon nanotube electronics will not be commercially viable unless carbon nanotubes are available in significant quantities and with the required quality and performance characteristics. Although nanomaterials firms are likely to be among the first nanotech firms to offer commercial products, and indeed, many of them are doing so already, they face a significant challenge in the not-too-distant future as the nanotech sector ramps up. It is one thing to be a small firm that provides very small quantities of a material, with very high gross margins to R&D facilities. It is quite another to produce specialist nanomaterials by the truckload with paper thin margins, while competing with the DuPonts and BASFs of this world, who wrote the book on how to make money in the specialty chemicals sector.

My guess is that as the market for nanomaterials expands many of today's small nanomaterials firms are going fall by the wayside because they simply won't be able to keep up with demand. Even if they offer the best quality, this is hardly going to matter if they can't support the volumes demanded by their customers. While it is going to take significant capital to ramp up to volume production, this may not be the biggest challenge in growing a nanomaterials business.[41]

Money isn't everything, and I suspect that moving from being a small firm that sells things in jars to scientists through a tiny sales force, to building a sales team capable of selling commodity-like products to large industrial enterprises is going to kill off more than a few budding nanomaterials firms within a few years. Either they will disappear altogether or they will simply be acquired by larger firms for their IP and engineering expertise. This is not actually a bad thing—it may well be exactly what the initial investors (certainly the venture capitalists among them) are after. Not all of today's emerging nanomaterials firms will disappear in this way, of course, and there may even be a few that find that they have a particular nanomaterial that is in such high demand that they can turn themselves into sizeable public companies. But remember that there are very few Microsofts, Ciscos, and Apples that can come from nowhere and become large market players in under a decade.

Nonetheless, nanomaterials remain central to the important nanotechnology as a commercial endeavor. It therefore seems to be of some importance to review the most important and pervasive of these materials and I do this, below. In addition, a final word on the distinction that I draw at the beginning of this section between nanomaterials and nanodevices; this distinction, while functional enough, is not quite as hard and fast as one might think. Nanomaterials are often smart materials, responding to changes in light, motion, pressure, and the like in a constant and intelligent way. In such cases, it may be a little difficult to decide whether we are talking a material or a device, in the sense that we are using these words.

Anyway, here is my Cook's Tour through the world of nanomaterials. It should be clearly understood that these are just a few examples of nanomaterials. Hopefully what follows will give the reader a good feel for the kind of thing that it is possible with nanoengineering of nanomaterials.

Buckyballs Also known as buckminsterfullerenes, or fullerenes, for short. Buckyballs are molecules consisting of 60 carbon atoms in a spherical formation. Commercial quantities of buckyballs can now be produced by burning hydrocarbon at low pressure and by other methods. Frontier Carbon (Mitsubishi) has built a factory capable of producing 40 tons of buckyballs per year.

Applications for buckyballs seem at present to be focused in the medical area. These include powerful antioxidants, drug delivery systems, and medical imaging systems. Other applications being commercialized include stronger plastics, more efficient fuel cell membranes, optical detectors, and coatings for computer disk drives. Major firms commercializing products based on buckyballs include DuPont, Exxon, Merck, Seagate, Sony, and Siemens.

Carbon Nanotubes Carbon nanotubes are well described by their name. They are nanoscale cylinders of carbon, with a specific atomic structure. This structure consists of a lattice of carbon atoms in which each atom is covalently bonded to three other atoms.

Like buckyballs, carbon nanotubes (CNTs) are a quintessential nanomaterials and as much of the rest of this book records, they are being commercialized by a wide variety of companies in an even wider variety of applications. This enthusiasm for carbon nanotubes comes from the fact that the capabilities of carbon nanotubes are quite remarkable. Indeed, if carbon nanotubes did not exist and some science fiction writer invented them, it is likely that such a writer would seem less a science fiction writer and more a writer of fantasies. Carbon nanotubes are an excellent exemplar of Arthur C. Clarke's comment (quoted at the beginning of Chapter 6) that any sufficiently advanced technology looks like magic. After all what can you say about a material that is all of the following things at the same time: stronger than steel, more conductive than copper, and one of the best known heat sinks known to mankind. An observer with less knowledge of the carbon atom than we do would certainly conclude that what we have here is magic indeed. (Table 2.1 summarizes some of the more important characteristics of carbon nanotubes.)

Actually, carbon nanotubes have quite a long history. They were first described in the technical literature in the late 1950s and a patent was applied for in the 1980s. However, the history of carbon nanotubes is usually measured from 1991, when Sumio Iijima, a researcher at NEC, created and photographed carbon nanotubes, explained what they were, and also provided them with a name.

Carbon nanotubes come in two species. There are single-walled nanotubes (SWNTs) that are simple cylinders. Multiwalled carbon nanotubes (MWNTs) are, as their name suggests, nanotubes each inside another nanotube. While SWNTs vary in size, on average they are about 1 nm in diameter and approximately 100 nm in length.

Table 2.1
Selected Properties of Carbon Nanotubes

Property	Comment
Strength	Greater than 50 times the strength of steel
Current density	~ 10^9 A/ cm^2
Elasticity	1 to 1.2 TPa
Thermal conductivity	More than twice that of a diamond
Density	About half that of aluminum
Chemical reactivity	Functionalizes like graphite
Thermal stability	Stable to 2,700° C
Chirality	Metallic or semiconducting

Source: Nantero.

Carbon nanotubes have been produced commercially for quite some time. However, until people started talking about the commercialization of nanotech a few years back, they were produced only in the quantities appropriate to a material used mostly in R&D labs. There is now a growing movement to produce CNTs in commercial quantities, although just what this means is not yet clear. Some of the initial commercial applications for carbon nanotubes will be in electronics, most notably in high-quality (field emission) television displays and in sensors. The performance and "unique selling propositions" of such devices will rely entirely on the special properties of carbon nanotubes, but individually they won't actually need a lot of carbon nanotubes. For example, the television displays mentioned above would use no more than just a "pinch" of CNTs and (theoretically) could be made to work with just a few individual CNTs. In practice, these displays would cost thousands of dollars, but use just a few dollars worth of CNTs. Where CNTs are used as a major component of some kind of fabric or other material that is used in bulk for (say) clothing, automotive bodies or some other similar application, they will have to be produced in much larger quantities. The reader should also note that the commercialization of nanotubes is almost certainly going to mean the emergence of various well-defined quality grades for carbon nanotubes. These are the tubes required to add in so-called bulk to a polymer to give it additional strength are not likely to need the same consistency of quality as those used for specialist electronics jobs.

Until quite recently, carbon nanotubes were produced primarily through deposition methods. One of the approaches used is high-pressure carbon monoxide deposition (HiPCO) This uses heated carbon monoxide gas that is broken up into carbon and oxygen atoms using iron as a catalyst. In the heated environment, the carbon atoms join with other carbon atoms to create the carbon lattice that is the nanotube. In chemical vapor deposition, the raw material from which the carbon is derived is methane gas or some other hydrocarbon. Again there is a catalyst, which this time may or may not be iron.

Newer and better ways of creating and manipulating carbon nanotubes are being worked on and could prove a source of competitive advantage for particular firms as particular ways of manufacturing certain chemicals have proved economically advantageous to certain chemical firms in the past. For example, a new plasma process for creating CNTs has been devised in which the heating element is a plasma torch, although the raw material is again methane. This approach has been claimed as being much more efficient than older methods in creating nanotubes and so is a step on the way to making nanotubes into a material that can be widely used in commercial applications. I should also mention that in applications, mainly in electronics and the life sciences, where carbon nanotubes must be placed in small numbers at a specific spot on a small area of substrates, new techniques will also evolve. These involve direct-write production approaches, such as DPN, that can place nanotubes in exactly the right spot.

Nanowires Nanowires are often seen as in competition with nanotubes, because they have very similar applications. However, in terms of physical structure and production technologies the two kinds of nanomaterial have very little in common. In particular, nanowires are not generally associated with any particular material. Nanotubes are almost always carbon nanotubes, although nanotubes have been built using other materials. Nanowires have been built using silicon, zinc oxide, indium oxide, and many other materials.

Nanowire technology is being developed by a number of research teams in universities and beyond. Nanosys, which is one of the most closely watched nanotech firms, has developed a nanowire technology platform that uses semiconducting nanowires for applications in the photovoltaic, display, and computer memory applications, and claims that nanowires are an improvement over what can be achieved in these applications using carbon nanotubes, because batches of nanowires can be manufactured with more consistent properties than for CNTs. Groups at NASA and at Hewlett-Packard are also working on nanomemory using nanowires. However, whatever the virtues of nanowires may be, it is hard for me to resist the impression that a lot more commercialization efforts are going into the carbon nanotube endeavor than into nanowire activities.

Nanocomposites In general, a composite is an engineered material made up of two or more materials. Composite materials are lighter, stronger and sometimes cheaper than noncomposite materials. There is nothing especially "nano" about the composites per se. Particle board used as a building material is a very good example of a composite, as is fiberglass.

From the comments that I have made so far, it is very easy to see how nanotechnology gets into the act. The extraordinary properties of carbon nanotubes, for example, make them an obvious candidate for use in new materials for electronics, building materials, automotive, and aerospace applications. There are good reasons for deploying these carbon nanotubes in the form of composite. Among these reasons is the tendency for carbon nanotubes to clump up, which can make pure nanotube fabrics inconsistent in density and hence weak in some areas. As a result of this phenomenon, Zyvex, for example, has created composites with nanotubes and polymers. Other reasons are the cost of bulk nanotubes, which would, at this point in time, make the cost of a "pure" nanotube material too high for widespread use.

Nanocomposites are already quite established in certain industries. General Motors has been using running boards constructed from nancomposites since 2001, as for instance, those found on the Safari and Astro minivans. The use of nanocomposites in this case adds strength to the machine, but also reduce weight and while a lighter running board is not going to make much difference to fuel consumption, once nanocomposites are more widely deployed in the materials used for both automobiles and for aircraft bodies and part, significant reductions in fuel use can be expected. This is, by the way, also an illustration of how the impact of nanotechology flows down to industry sectors, from one of the big three areas. In this case, the use of nanomaterials could well be classified under the nanonergy label.

Nanotechnology and Smart Materials The nanomaterials that are profiled briefly above are merely illustrative of the many kinds of nanomaterials that will soon be available commercially. There are nanocoatings, nanopowders, nanocrystals, nanofibers, and even nanopastes. All contain remarkable properties. To do all of these wonderful new materials justice would take a book dedicated to this sole purpose and it would be a long book. Faced with the prospect of the commercialization of many new materials, materials and chemical firms will be forced to ask themselves some classic questions for their industries. There are inevitable questions, such as will these new materials hurt the market for my existing products? And the classic question: if my business becomes involved with the new nanomaterials, just how much of the value chain do we want to become involved with?

Different firms will have different answers to these questions. In particular, their answer to second question may mean rethinking their entire business plan. For example, I worked on a project assessing the market potential for new materials that can be used for RFIDs and similar products. In talking with firms that are actual or potential manufacturers of such materials, I found that most of them simply saw RFIDs as a high potential market for their materials because they are likely to gradually replace barcodes in many instances and would therefore be created in the millions or even billions. However, there was one firm I encountered who believed that the only way to really squeeze value out of the growth of the RFID market was to get into the RFID manufacturing business itself (albeit with a partner). In other words, the nature of this opportunity was such that in order to best capitalize on it, the firm in question had to fundamentally alter its business strategy.

Although the example given above is highly unusual at the present time, I believe that in the future new nanomaterials may inherently push manufacturers up the value chain. This is because many of the materials that nanotechnology will enable will be "smart" materials. These are usually defined as those that can sense external stimuli and then respond to those stimuli by some means in real (or near real) time. Such materials are therefore a product that lies somewhere between a material and more complex nanodevice—a sensor, in fact. Some smart materials respond to light, others to pressure, and so on. Some are self-repairing. The idea behind the evil Terminator in the movie *Terminator 2*—the one that kept getting up and rebuilding itself whenever it was destroyed by Schwartzenegger—came from developments in the area of shape metal alloys (SMAs), a particular kind of smart material.

While there are real SMAs that are being marketed now, none have properties quite as spectacular as this. In fact smart materials have been talked about and created in a limited way for decades, but the advent of nanotechnology and other developments in materials science are gradually making them a product that some firms are seeing as likely to have a significant commercial impact in the not-too-distant future. Areas where they may be deployed would include large area sensing arrays, robotics (for robotic skin, for example), clothing, and military uniforms (where they would respond to changing environments), the bodies of many different kinds of vehicle (from cars to space ships), regenerative medicine and even in information technology (where they seem to have some potential to create denser disk drives).

In many of these areas, smart materials promise significant cost reductions and even more significant performance enhancements. And there are a host of technologies that are being used (or considered) to create them. These include well-established technology such as piezoelectrics, MEMS, and a variety of nanomaterials including buckyballs and nanocomposities. However, as a firm moves from being a materials company to being a smart materials company, it also faces some strategic challenges. Suddenly it has also changed from being a firm that sells what is more or less a commodity to one that is in effect selling a kind of sensor product, with all the concerns that firms higher up the value chain currently have. This is one of the many examples of how nanotechnology shifts traditional industry boundaries that you will find throughout this book.

A Word or Two About MEMS

If we have less to say about nanomaterials than we do about nanodevices and nanosystems, we have even less to say about microelectrical mechanical systems (MEMS), which is a technology that is often put together in a clump with nanotechnology. MEMS are tiny machines, built using conventional silicon semiconductor manufacturing processes, often in old semiconductor fabs. MEMS devices have found a myriad uses as components in airbags, projection televisions, optical switches, sensors, and many other areas. MEMS involve microtechnology, basically a specialized form of computer chip and not in any sense nano.

Nonetheless, nanotechnology and MEMS are often discussed together and sometimes they are both included under the same heading of "small tech." The assumption behind this categorization appears to be that MEMS is the big brother to nanotech. MEMS comes first and then the small tech sector would start to shrink devices down to nano size. It is certainly true that nanotechnology will continue the path blazed by microtechnology towards miniaturization and will lead to some of the same economic consequences, in particular continued decreases in the price/performance ratio of many products. However, in my view, this is where the similarity ends, except in the case of one particular kind of nanotechnology called NEMs, which really is the nanotechnological (i.e., further miniaturized) version of MEMS, but happens to be one of the least developed nanotechnologies at the present time in commercial terms. It should also be noted that MEMS and nanotechnology are sometimes combined in products based on AFM technology, as for example, in the certain kinds of nanomemory device and dip-pen nanolithography.

Even though this is not a book on MEMS, the fact that nanotech is so frequently considered along with MEMS seems to call for a brief discussion of why and how the two technologies have become intertwined. As far as I can tell the main reason seems to be history. Eric Drexler's original view of nanotechnology was one in which very small machines performed numerous useful tasks. Although these were molecular machines and MEMS are basically just computer chips operating at the micro-level and mostly under the laws of classical physics,[42] MEMS has seemed to be the nearest thing to Drexlerian nanotechnology that has yet to become a real product. Hence the association that was, and continues to be, made.

Sometimes there is a connection between nanotech and MEMS that goes beyond mere metaphor. In some cases, MEMS devices are considered to be interim devices that will eventually give way to nanodevices. This is true of sensors, for example, where there is much talk these days about "smart dust," a kind of very low-cost sensor that can be therefore widely distributed in order to provide timely and geographically specific information where this is required, such as in homeland security, military, and meteorological applications. In practice, despite the name, "smart dust," units aren't actually that small or low cost, and, at the present time, they are mostly MEMS-based devices. It is easy to see how a direct transition from a MEMS-based system with further miniaturization to a NEMS-based system may enhance the value of smart dust.

Beyond such applications, however, speaking of MEMS and nanotech in the same breadth can be confusing for a number of different reasons:

Nanotech Is Bigger Than MEMS While MEMS technology really is closely related to NEMS technology, one being a further miniaturized version of the other, there is no connection worth mentioning between MEMS and many other kinds of nanotechnology. For example, MEMS and nanoengineered photovoltaic cells are about as closely related as the technology of a diesel engine and that of water-soluble paint. As the example of smart dust indicates, there may be a simple transition from MEMS to NEMS. There is, however, no reason to assume that such a change may occur in all applications. While MEMS devices are small, they are by no means leading edge in this regard from the perspective of the semiconductor industry. Leading edge processors have much smaller features.

Nanotech Is Newer Than MEMS MEMS is a well-defined, well-established industry containing profitable firms that already ship large quantities of products every year. MEMS is most commonly found in automobile air bags and in DLP[43] televisions. Nanotech products are a testimony to, and are still, almost invariably, in their early stage of development. Hence the business and financial models in nanotech and MEMS are typically very different.

To summarize, the replacement for a MEMS product may be a NEMS product; or a MEMS system may be replaced by a nanosystem or nanodevice that is entirely different from the MEMS system in terms of material and architecture; or a MEMS product may not be replaced at all. There may be much that a nanoengineer can learn from a MEMS engineer, but for the most part the technologies are quite different and the term "small tech," seems somewhat ineffective. The engineer developing the latest MEMS device for an optical switch has little to share with a nanomaterials expert developing the latest type of nanowires, except the time of day.

Four Types of Nanotech Business Opportunities

Hopefully, I have said enough so far to indicate the great commercial potential of nanotechnology and why people are getting so excited about it. In a nutshell what people are sensing is that here is an opportunity to create entirely new products, even, perhaps, a whole new segment of the economy. If nanotechnology never turns out to be an epoch-making technology in the sense that I have defined it earlier, and even if there is never a nanotech boom or bubble there is going to be an accelerating number of nanotech business opportunities that will emerge in the coming years. You can be certain of that. However, there will be an equally broad range of risk/return profiles associated with these opportunities.

Working with clients in this space, it appears to me that despite the many different kinds of nano-enabled businesses that are likely to appear in the next decade, they can all be classified into four different types. We will refer to them as accidental nanotech, evolutionary nanotech, revolutionary nanotech, and disruptive nanotech, described as follows:

Accidental Nanotech As I indicated in the previous chapter, some uses of nanotechnology are really quite old and are part of traditional businesses. I will refer to them as accidental nanotech, and this term covers the kind of nanotech opportunities that would never get anyone very excited. They just happen to qualify under the current definition of nanotech. No one would ever write a book about them.

A good example of accidental nanotech is the use of carbon black for tires. There is no doubt that this is commercial nanotech, but it is hardly much of a new business opportunity. Indeed, nobody who isn't already producing carbon black is going to enter such a business unless they have some new and highly profitable way of using carbon black. There are, of course, many firms already selling carbon black and for the most part they are not going to quit the business unless tires start being made in an entirely different way.

The bottom line is that accidental nanotech opportunities aren't even really opportunities in the strict sense, but for established firms they represent relatively low risk ways of generating revenues. The biggest barriers to entry are likely to be the low profitability that they offer. Firms that are involved with incidental nanotech are often the very antithesis of the kind of high-tech firm that one most naturally associates with nanotech.

Evolutionary Nanotech Evolutionary nanotechnology is really a fairly straight extension of materials technology with some control being applied at the nanolevel. In other words it is not as accidental as true accidental nanotechnology. Nonetheless, the products of accidental nanotechnology are not the kind of thing that gets people very excited: items such as stain resistant pants and nano-enabled cosmetics.

In a sense, evolutionary nanotechnology is a proof of concept for the complex nanoengineered products. They are available now and can be seen as testing the water for the even more complex nano-enabled products that are yet to come. Evolutionary nanotechnologies enable new products and market segments to be created that have sufficient novelty and potential profitability to attract new business, but everyone knows (or should know) that they aren't going to change the world. There are some real risks though—what if nanoengineered cosmetics leave the buyer unimpressed?

Revolutionary Nanotechnology Revolutionary technology pushes up the risks and returns one more notch and is the kind of nanotechnology with which this book is principally concerned. With revolutionary technology we move beyond products that have been around forever, but just happen to fall into the definition of nanotech and also beyond products that are interesting new variations on older products. Here we have genuinely new products:

- In the nanoelectronics area, I am talking about very fast, high-capacity nonvolatile nanomemories that would enable HDTV movies to be stored and played from an iPod-like player. I am talking about flexible displays that can be rolled up and put in your pocket when they are not needed. Or the nano version of smart dust sensors.
- In nanomedicine, revolutionary products might include a nanoengineered capsule that would be injected intravenously and would be capable of burning away cancer cells. Anyone who has ever had a loved one endure chemotherapy will understand why this would be a revolutionary development. Other revolutionary products in the nano- medical space will include a broad range of replacement parts, especially artificial skin and artificial joints. The unique selling proposition of nanotechnology in the medical space is that it operates at the molecular level, the same level at biology operates and so nanodevices seem intrinsically better at functioning and blending in than the alternatives. Some critics of nanotechnology, however, see this as precisely the problem with nanotechnology.
- In the nanoenergy sector, revolutionary products would include nanocatalysts or other additives that doubled the efficiency of the fuel consumption of vehicles or which made the all-electric car a near-term possibility. Yet another would be carbon nanotube based electricity transmission systems that enabled truly distributed power grids. Unfortunately, none of these seem likely near-term products. A more likely possibility for a revolutionary energy product that will generate some revenues in the near term comes from the photovoltaics sector. Here a small group of firms both large and small—Konarka and Nanosolar on

> the one hand, and British Petroleum and General Electric on the other—are working on low-cost photovoltaic cells, based on nanomaterials. The potential of such products lies in powering mobile computing and communications devices, in pushing the geographical limits in which PV is an economically viable option for providing power to buildings, and in creating smart materials that are self-powering.

Revolutionary nanotechnology will mean a serious rethinking of marketing strategies by many companies as they deal with that riskiest of propositions, that of selling entirely new products, sometimes into entirely new markets. In this set of circumstances there are bound to be significant failures, but there are also bound to be significant successes. It is revolutionary nanotechnology that seems most likely to produce the Intels, Microsofts, and Ciscos of the future and many of the first nanomillionaires. One may assume that many of the large public nanotech firms of the next decade have probably not even filed for registration yet.

Disruptive Nanotechnology Truly disruptive nanotechnology, in the sense that I am using it,[44] is an area to which we will devote relatively little space. What we are talking about here is nanotechnology in the Drexlerian mode with the whole of the world economy gradually remade in the nanotech image. Even its strongest proponents say that it could take a few decades before this kind of technology evolves.

Disruptive nanotechnology is not an area where real business opportunities are likely to be found in the foreseeable future, although this does not (and should not) stop researchers working in this field. In the end, it may be the disruptive nanotechnology that generate the biggest revenues of any of the different kinds of opportunities listed here. This will not occur until a long time in the future and these are not the kinds of opportunities that it will be easy to convince investors and corporate funding committees to put money into.

The Importance of IP in Nanotech

Intellectual property (IP) is a term that is used for patents and copyrights, but will be used here more broadly to include the ideas behind a business or product and not all of these ideas will be protectable in a formal legal way. As I have mentioned, IP-based business models have been in favor in the nanotech community for start-ups, although, as I will now discuss, there are reasons to be quite skeptical of them.

That said, at the present time it is hard to overestimate the importance that investors and the nanotech community places on IP. If you visit almost any major nanotech trade show or conference, you will find yourself bumping into numerous IP lawyers and other IP-related professionals. At the trade shows, you should expect to see quite a few booths established by prominent IP law firms, which is a little surprising, in some ways. Having attended numerous trade shows in quite a few disparate industry sectors over the years, nanotech shows are the only ones I can think of where IP attorneys have big booths.

As with any other business, considerable weight must be attached to the uniqueness behind a nanoproduct and the degree to which those ideas can be protected from being used or circumvented by other people. However, I am personally skeptical about the degree to which IP will continue to get the attention that it now gets in the nanotech community. Some of the reasons that I believe this are simply commonsense business reasons that might apply to any high-technology business:

- In the end, the business is going to succeed or fail to a very large degree on the quality of its technical, marketing, financial, production, and general management, much more so than its IP. Indeed, for nanofirms management may be especially challenging because managers must also be competent (and even better, excellent) technologists and scientists, if they are not to make extravagant promises (or be hoodwinked themselves) about what their products can and can't do.
- Basing a strategy on a pure IP model, in which ideas are generated and then licensed or sold off may be something of a missed opportunity. Where the ideas are genuinely novel, there may be as much money or more to be made from technology transfer fees as from licensing itself. In simple terms, licensees may be prepared to pay for consulting on how best to implement the technologies they are licensing. Such arrangements may well develop into lucrative product development, manufacturing, and marketing partnerships. In any case, while the pure IP model seems, in theory, to avoid having to make the hard strategic choice of being in one business or another, the reality is often somewhat different. For one thing, investors (especially VCs) seldom want to see the firms in which they are invested spread too thin. In addition, nanotechnology platforms are almost always complex to the point where an IP-oriented firm can seldom just sell an idea and be done with

it. It must also provide a technology transfer package and ongoing support to its customers.

Is the Role of IP in Nanotech Exaggerated? Contrary to the beliefs of many, I think that the pure IP model may ultimately prove to be especially hard to defend in the nanotechnology business. This is because nanotechnology provides a very wide range of materials and manufacturing platforms. This, in turn, suggests that performance goals for nanoproducts can be achieved by very different routes. We are already beginning to see this occur in a number of areas of nanobusiness. For example, in the search for high-capacity, nonvolatile computer memory, a wide variety of completely different nanotechnologies are being considered, as for example, thin-film magnetics for MRAM, organic electronics for polymer memory, and carbon nanotubes for carbon nanotube memory. Even if a firm could achieve a great leap forward in commercial memory technology using one approach and could file a foolproof patent for its technology—which is a somewhat unlikely event—there is nothing that a patent could do to stop it being bested in the marketplace by another memory approach. Nanotechnology makes such an event a lot more likely to occur than in the days when the computer chip business was all about silicon electronics, since it provides new materials from which to construct memories and hence a whole new way of getting around patents.

If my assessment of the real role of intellectual property in the future of nanotech turns out to be correct, then it may have some interesting consequences for how nanobusiness takes shape. First, there has been a tendency to regard protected IP as a key factor in the valuation of start-ups. The importance of this factor may be exaggerated. With this in mind, it is an important strategic decision that must be taken by management as to the degree to which it makes sense to become involved with manufacturing or remain a largely an IP shop.

There is always going to be a trade-off between just selling ideas with little capital input to the business and going whole hog and installing a plant. In the semiconductor industry where leading edge plants (fabs) now cost billions of dollars and can be afforded by just a handful of manufacturers, the sheer impossibility of being able to afford to manufacture by most firms has led to the creation of two entirely new industry sectors: fabless chip suppliers and the foundries that manufacture the chips for those suppliers. However, while not trivial, the capital investments made by nanotech firms are not at a point (as they are for the mainstream semiconductor industry) where they are simply beyond the capabilities of all but the very largest firms. So at this point, most firms have a genuine choice about what role IP will play in their business model.

This is not to say that IP can ever be ignored and protection of IP will always be an issue to which much attention will be paid.

Short-Term Concerns About Nanotech IP In addition to these strategic concerns about IP in nanotech, there are also short-term issues that need to be dealt with about nanotech patents.

These problems are ones of success, as indicated in the sudden surge of nanotech patent filings. According to one source, as of late March 2005, 3,818 U.S. nanotechnology patents had been issued with another 1,717 patent applications awaiting judgment.[45] A similar surge in all the developed nations has strained the capabilities of patent offices worldwide to cope with demand. Nonetheless, this is felt most acutely by the U.S. Patent and Trademark office, because of the concentration of nanotech firms in the United States and because IP protection is given so much weight in the business models of U.S. firms. The basic problem is easy enough to understand. There are too few examiners without the technical training required to review nanotechnology patents. The likely results will be invalid claims, overlapping claims and claims that are far too broad, not to mention claims that are valid but are, in the first instance at least, rejected by the PTO.

Such problems seem likely to become prevalent worldwide and will ultimately have to be addressed with legislation and, most likely, higher budgets. For now, however, and for quite some time to come, the nanotech patent thicket goes a long way to explain all those patent attorneys that keep showing up at nanotech trade shows.

The National Nanotechnology Initiative and Other Government Nanotech Programs

Government intervention to improve the nanotech patent process seems at this point like a very reasonable expectation. Indeed, many governments around the world are convinced that the development of nanotech more generally is in need of funding out of the public purse and steering at the governmental level. In 2004, the world total of government funding amounted to $4.5 billion, compared with $1.3 billion in 2001.[46] The biggest programs are the United States government's National Nanotechnology Initiative (NNI) and Japan's government program, which is similar in size and scope to the U.S. program, but with, perhaps obviously, a somewhat Japanese flavor. Other nations are going down the same path, but typically with lower levels of funding. In total, government funding for nanotech in the North American, European, and Asian-Pacific regions are roughly equal.

This section describes what is going on worldwide in terms of government support. I have given a bit more attention to the NNI, because it seems to have become a model for government nanotech programs more generally.

The NNI The NNI is run by the National Science and Technology Council, whose members are appointed by the President of the United States. It does several things. It is designed to coordinate nanotech related efforts throughout many different government agencies. It provides funding for nanotech R&D projects and more generally is expected to do a general outreach program to get businesses, both large and small, to become more involved in nanotech, especially through university/industry partnerships.

The participants in the NNI comprise a wide variety of U.S. government agencies, including the Departments of Defense, Energy, Justice, Transportation, Agriculture, State, and Treasury. Also involved are the Environmental Protection Agency, NASA, the National Institutes of Health, the National Institute of Standards and Technology, the National Science Foundation, the Nuclear Regulatory Commission, the CIA, and various groups within the Executive Branch (i.e., the White House). In short, there are few areas of government activity that are not involved in the NNI and each agency has a high degree of flexibility to allocate funds to support its own goals, including the government funding programs for nanotech described elsewhere in this chapter. More generally, the implementation of the NNI is centered around seven different kinds of programs and four important goals. The programs and goals of the NNI are listed in Table 2.2.

Not all of what is set out in Table 2.2 has that much to do with commercial matters, which, of course, are the primary concern of this book. However, some of them do and seem to indicate where the NNI will have its greatest positive commercial impact. Understandably, the program focusing on nanomaterials is seen as critical to the goal of the NNI in supporting R&D aimed at realizing the full potential of nanotechnology. Interestingly, however, when it comes to economic growth and development, the programs that are seen as critical in the NNI are the ones that involve devices and systems. This seems to confirm one of the major premises of this book, which is that the long-term business opportunities stemming from nanotech will come from the more complex nanoproducts, not so much from the nanomaterials themselves. For more details on U.S. federal government activity in the nanotechnology space, see www.nano.gov.

Table 2.2
Programs and Goals of the NNI

Programs	Goals: R&D aimed at realizing the full potential of nanotech	Facilitating transfer of new technologies into products to promote economic growth, jobs, and other "public benefits"	Developing educational resources, a skilled workforce, and the supporting infrastructure and tools to advance nanotechnology	Support responsible development in nanotechnology
Fundamental nanoscale phenomena and processes	C			
Nanomaterials	C			
Nanoscale devices and systems	P	C		
Instrumentation, metrology, and standards		C		
Nanomanufacturing		C	P	P
Major research facilities and instrumentation acquisition	P	P	C	
Societal dimensions			P	C

Source: NNI.
C = Program is critical to goal.
P = Program has primary relevance to goal.

Other U.S. Programs The NNI is 400-pound gorilla as far as U.S. nanotech programs go and its preeminent role has been confirmed with the signing into law by President George W. Bush of the 21st Century Nanotechnology Research and Development Act, which authorizes funding appropriations at federal agencies for nanotechnology R&D programs.

However, there are other important activities in the U.S., that are designed to promote nanotech activity. At the federal level, the National Science Foundation, through the National Science Board, has some activity independent of the NNI. There are also very substantial state programs in several U.S. states. These are reviewed in Table 2.3.

Europe At every level Europe is struggling to present itself as a single political entity and its nanotech work is being coordinated through the European Commission, which is the pan-European bureaucracy. Funding for nanotech is to be found primarily in Framework Program 6 (FP6), which is budgeted to run from 2002 to 2006, although FP6 covers a number of other areas too. Detailed profiles of current activities and funding can be found at the (European) Community Research & Development Information Service Web site (www.cordis.lu/nanotechnology).

In addition to the pan-European activities there are also numerous initiatives at the level of the individual nations. These programs provide important sources of funding for R&D projects in nanotech, but are in many ways political animals subject to all the usual twists and turns that accompany any essentially government program. One group that provides regularly updated reports on the state of European and other national nanotech initiatives is The Institute of Nanotechnology (www.nano.org.uk).

Japan Japan has a long history of industrial policy, a fact that derives in part from its historical goal of catching up with the West, an objective it can claim to have achieved in every meaningful way. The pattern of Japan's industrial policy has usually been to combine government funding with a collaborative program between the very largest Japanese industrial conglomerates and various government agencies.

This pattern can be discerned in current efforts in the nanotech field, which do not appear to be giving much priority to the activities of small- or medium-sized firms. As a result, the government collaboration is likely to involve Japan's largest chemical, electronics, and pharmaceutical firms. The two main government players are the Ministry of Education, Culture, Sports, Science and Technology and the Ministry of Economy, Trade, and Industry. This reflects the twin goals of Japanese nanotech policy which is to promote nanotech and nanoscience research in prestigious universities, such Tokyo University and Kyoto University *and* the importance that Japan's government places on the economic development role of nanotech.

Table 2.3
U.S. State Programs in Nanotechnology

State	Nature of Program	URL
California	The Northern California Technology Initiative (NCNano) has very big nano plans for the Silicon Valley area and says that it will attract billions of dollars in nano investment and R&D funding and as many as a 150,000 jobs.	http://www.ncnano.org
Colorado	Colorado Nanotechnology Initiative provides support for education and public awareness projects	http://www.colo-rado.nano
Connecticut	The Connecticut Nanotechnology Initiative is a collaborative arrangement between universities, private industry, and government agencies.	http://www.ctnano.org
Massachusetts	Massachusetts Nanotechnology Initiative is a project of the Massachusetts Technology Collaborative (MTC). Goal is to support R&D and new ventures.	The MTC's URL is http://www.mtpc.org
Minnesota	Plans for opening a nanotech research center at the University of Minnesota. The research center will be called the Organization for Minnesota Nanotechnology Initiatives, which just happens to have OMNI as an acronym.	http://www.nano.umn.edu/omni
New Jersey	The New Jersey Nanotechnology Consortium is centered on major research facilities in New Jersey including Princeton and Rutgers, as well as Bell Labs. Its goal is the commercialization of nanotechnology.	http://www.njnano.org
New York	Large state/industrial program to create R&D centers along with some venture capital funding. Focus seems to be broader than nanotechnology including photonics, bioinformatics, IT, and environmental technology. Within nanotechnology, Albany Nanotech, which is located on SUNY, facilities has become a force to be reckoned with in nanoelectronics.	http://www.albanynanotech.com
Texas	Texas Nanotechnology Initiative is a consortium of universities, private industry, and government organizations to promote nanotechnology in Texas.	http://www.texasnano.org
Virginia	The Initiative for Nanotechnology in Virginia is a consortium of universities, state agencies, federal labs and industrial partners. Its goal is to promote economic development, R&D, and commercialization.	http://www.inanoVA.org

In addition to government ministries a variety of national research institutes are involved with Japan's national nanotech efforts. These include the National Institute for Materials Science (www.nims.go.jp/eng) and the National Institute of Advanced Industrial Science and Technology (www.aist.go.jp/index.en.html). Japan is also the venue for the world's largest nanotech conference and trade show.

China As few reading this book will need to reminded, China has become the world center of manufacturing. Nanotech, of course, has much to with manufacturing, so it is no surprise to see that China has a government program. Most of the activities seem to be focused on a couple of specialist research institutes—the National Center for Nano Science and Technology (NCNST) and the National Engineering Research Center for Nanotechnology (NERCN). There is also a Shanghai National Engineering Research Center for Nanotechnology (SNERC). The focus of much of the nanotech R&D being carried out in China at the present time is nanomaterials, and especially the mass production of nanomaterials. The other area that seems to be of some interest in China is that of nanosensors for both healthcare and IT applications.

Ted Fishman, in his best-selling book on the economic rise of China, *China, Inc.,* specifically mentions nanotechnology as an area where China has little competitive advantage and, at the present time,[47] China is certainly not a force to be reckoned with in the world of nanotech and their government investment in the area is fairly small. I have been told the total amount is in the region of $50 million annually at the present time, although as the huge advantage that China currently has in terms of labor costs begins to ebb away, we can be certain that the country will start to turn to advanced technologies as a way to ensure its competitive advantage. We can also be sure that technologies with a military perspective to them will occupy the minds of Chinese leaders. Nanotechnology fits into both categories.

India Another country that is now emerging as an economic powerhouse is India. While China is seen as a manufacturing center, India is generally viewed more as a source of low-cost service workers, whether that service is call center staffing or software development. This role has been promoted by the fact that English is widely spoken and the country has long had good technical education at the university level and above.

The Indian National Nanoscience and Technology Initiative (NSTI) emphasizes all the usual suspects—research, education, and commercialization.

Just as in China, there is a heavy emphasis on nanomaterials, although there is also interesting work going on in nanoelectronics.

Israel In the past decade or so, Israel has emerged as an important source of high-technology R&D and commercialization, with much of the origins of this work apparently finding its origins in Israel's large military sector. Israeli start-ups and venture capital firms are modeled after those in the United States and so is the Israeli Nanotechnology Initiative (INNI).

The group within the INNI that raises and distributes money is the Israel Nanotechnology Trust (http://www.nanotrust.org.il) There is also a group in Israel called the Consortium for Nanofunctional Materials that is a collaboration between private industry and academia. While Israel may never be able to match the clout of the United States, the Israelis have been surprisingly influential in the development of new IT technology and are strong in the biotechnology and semiconductor sectors. It would not therefore be surprising to see this country do well in nanoelectronics and in nanobiotechnology. Other areas where Israel seems likely to be influential, because of its existing expertise, is in areas where nanotechnology can contribute to water desalinization and photovoltaics.

What to Expect from Government Programs to Nanotech This is not a book about policy or politics. However, a reasonable question for any businessperson to ask is what are the likely outcomes of government programs in nanotech? In particular, will these programs really jump-start nanobusiness in the way that their supporters hope? There are clearly many people in the United States who believe that nanotech could never take off commercially in this country without the NNI and it seems likely that the importance of government funding is ranked even higher in Europe and Asia, where industrial policy and government/business collaborations are more accepted than they are in the United States.

A number of arguments can be brought to bear on the side of the need for government programs to pump prime nanobusiness:

- Many major civilian technologies appear to have got their start through government funded projects. This could most clearly be argued about advanced communications technology, but is even true of the transistor to some extent.
- In the United States it has been argued that the NNI was necessary to turn the tide from nanotech being something molded in the manner of the original Drexlerian vision of molecular engineering to something

more. People who take this point of view believe that the government's embracing of nanotech as something more practical than molecular engineering is the best hope of moving nanotech forward.

For better or worse, there is something of a consensus in the nanotech community that without significant government funding, nanotech will not make much progress. This is why the nanotech community has given four cheers for the NNI in the United States. It is also why groups from various segments of the nanotech community continue to lobby for more government funding either for nanotech in general or for specific products. For example, the late Richard Smalley, the inventor of the buckyball, has called for government funding along the lines of President Kennedy's space program to use nanotechnologies to make the United States less dependent on fossil fuels.

There are also some voices that oppose this kind of funding. C. Wayne Crews of the Competitive Enterprise Institute has made the very plausible point that government funding now means more regulation in the future.[48] In other words, he who pays the piper is he who calls the tune. Such voices are in the minority right now, but, I suspect will become more common, if some of the current government programs fall far short of expectations.

Whatever the historical reality of government involvement in the earliest days of a new and important technology and whatever the legitimacy of its involvement in funding the basic science that makes that technology possible, there are real questions about whether big programs designed to designed to promote commercialization of technologies are very effective. I have written fairly extensively on how ineffectual such programs have been in telecommunications in my book, *Telecompetition.*[49] In general the history of ambitious government projects designed to pick the best technologies and then back them with big bucks, has been, not to put too fine a point on it, horrible. Who can forget the French government's attempt in the 1980s to jump-start the French cable and informatics industries, or the Japanese government's effort to produce a new breed of artificial intelligence device at about the same time, its so-called Fifth Generation Project? The answer seems to be, just about everybody.

It has been suggested that, at least in the case of Japanese program, there was never any serious intention to reach the stated objective, but rather, the program was a way to create a rallying cry. If this is the case, it may have been a rather expensive PR campaign. In the United States, government funded science programs are sometimes justified in terms of spin-offs and it is true that defense related technology often does have important commercial spin-offs. But this is hardly the point of defense spending. After the first blush of excitement about the U.S. space program, the high expenditure on the program was often justified in terms of spin-offs, although this was always very hard to prove. In fact, what seems more likely is that by luring industry to make investments in products that the government believes are priorities, large government R&D programs simply draw the energy of private enterprise away from what the marketplace actually wants.

Anyway, this is not a book on nanotech policy, so I will take this argument no further. However, the key takeaway from all this for the practical business-person is that if you are looking to a whole new nanotech sector to emerge because of government programs, you may be sadly disappointed.

Funding Issues for Nanotech Businesses and Projects

The sources of finance for a nanotechnology business are ultimately the same as any other high-technology business. That is, the smallest most entrepreneurial firms will finance their activities from angel investors and the personal funds of management and their relatives. Somewhat larger firms will get their funding from investment banks and venture capitalists. And the nanotech projects at the largest multinationals are being funded from internal cash flow and the great public capital markets. I suspect that internal funding of corporate nanotech projects are going to account for the bulk of actual revenues from nanotechnology in the foreseeable future. As in so many high-tech sectors, small VC-backed firms often have fascinating ideas, but simply lack the clout in the marketplace to turn them into business realities.

Especially for these smaller firms, there are, however, some special circumstances that must be taken into consideration when funding a nanotech firm, particularly at an early stage. One of these is that, as mentioned earlier, nanotech start-ups typically need a significant amount of capital compared with a dot-com. Nanotech also differs from some high-tech sectors in being more likely to get money from the government. In the United States, there are a number of government sources of funds, both federal and state. The federal sources are reviewed in Table 2.4.

Government funding is often seen by firms as a form of free money. This has some truth to it, but, even though firms are typically allowed to own the IP they have developed with government money, there are real costs associated with this government funding. These costs include the following:

- *Product strategy cooption by government requirements.* Product strategies must be adjusted to meet the funding requirements of government

Table 2.4
Sources of U.S. Government Funding for Nanotech

Type of funding	Agency	Applicability
SBIR grants	Several agencies	Only for small businesses. This is a three-phase program with up to $100,000 for feasibility studies in Phase I and a two-year development grant of up to $750,000. Phase III involves private funding and seldom is used.
STTR grants	Several agencies	Similar to SBIR grants, but require that researchers at universities and research institutes play a major role in the project.
DARPA funding	Department of Defense	Oriented towards potentially high payoff research of importance to the military.
Advanced Technology Program	NIST	The goal is to provide matching funds for up to 50 percent of the cost of R&D for "high risk, high potential" technologies developed by industry. Money can go to firms of any size, and about half of the grants go to smaller firms. Grants can be quite large compared with other programs, but government gets a share of the royalties and licensee fees from ATP funded projects.

grants. This is the reason that so many firms who hold the IP on nanotech technology platforms that could be used for a wide variety of products are focusing on products such as nanosensors that can be used for military and homeland security applications. (An interesting question, which I only intend to pose here — not answer — is whether the orientation towards government funding therefore actually ends up delaying commercialization of nanoproducts aimed at the business and consumer markets.)

- *The costs bureaucracy.* Applying for government grant to provide the initial funding for a nano enterprise is exactly what you might expect from a government program—lengthy and bureaucratic.
- *The costs of competition and politics.* Significant amounts of money can be spent by small nanotech firms trying to obtain government grants, only to find that they never stood much of a chance in the first place.

> Precisely because government money is seen as a source of free money, the process can be very competitive. In the end the money is likely to go to firms whose technical staff have the most respectable resumes and/or the best political connections.

However, despite these limitations, winning government grants is probably the cheapest source of capital available for U.S. nanotech firms and it tends to impress outside investors. This is certainly the case with venture capitalists and is probably the case for angel investors too. Angels are, in fact, another popular source of early stage funding for nanotech firms and angel investor networks in California and New York have specifically been created to invest in nanotech opportunities and angels are often well suited to the needs of the high-risk early stage business that constitute the current nanotech sector.

But it is the venture capitalists, not the angels, who are probably the source of finance that many nanoentrepreneurs think of first. However, VCs are really only an appropriate source of finance for the nanotech firms with revolutionary products, because it is only they that have the potential for capital growth that venture capitalists require to satisfy *their* investors. Hence many nanotech opportunities are indeed genuine opportunities. They are just not the kind of opportunity that a VC would finance. Indeed, VCs are looking for more than just potentially high returns. Other factors that influence them when the offer money are:

- Whether the firm already generating some revenues.
- Whether the firm has an established management team that already has some successes behind it. One of the reasons why people got so excited about the prospects for a Nanosys IPO is that the management team behind it had already proved its worth in the biotech sector.
- Whether the science behind the firm has a good pedigree. This would be established if the firm grew out of efforts at one of the major universities, industrial labs or a national lab. If a famous nanoscientist is a key part of a firm, this would certainly be an incentive to many VCs to give money.

Finally, many VCs have not been especially friendly to nanotech and have been mistrustful of it as a sector. Venture capitalists are not rushing to fund nanotech firms by the droves. One reason for this is that the materials/technology platforms that nanotech firms claim for themselves often have so many applications, outside of the pure IP model, that it becomes quite difficult to pin down precisely the sources of the revenue. While some might take the view that this situation represents and embarrassment of riches, this is not the view that VCs are likely to take and one of the conditions that they are likely to insist on is that, whatever the potential of a firm's technology, it focuses closely on just one or two products in an equally narrow range of applications.

Today, there are just a handful of VCs that are really emphasizing nanotech in their portfolios. These include a handful of firms that have made nanotech a special focus. Such firms include Ardesta, Draper Fisher Jurvetson, and Harris & Harris. There are also a few of the best-known VCs who have been prepared to make nanotech a special focus. These include ARCH Ventures, Morgenthaler Ventures, and Sevin Rosen Funds.

Nanotech, Safety, and the Environment: Nanotech's Little PR Problem

Back in 2004 I was a speaker at the Nanocommerce trade show and conference that is held in Chicago every year. As is the case with most such shows, there are receptions held most evenings that enable speakers, attendees, and exhibitors to mix and meet. These are not the kinds of events that usually get gate crashed. They are typically a little too dull for this. Evening events at nanotech conferences have never, as far as I know, been favorably compared to a Hollywood party.

But the 2004 Nanocommerce show was different. In the middle of the evening attendees at the party were more than a bit surprised to find that the party had not only been gate crashed, but that the gate crashers were naked. The nudity it turned out was inspired by a protest against nanotech, which the protesters believed was harmful to humanity.

It is easy to criticize such protesters for their naivety (and their nudity), but as nanotech becomes more visible to the public this kind of things is going to become more common and will become a real PR issue for nanotech firms, of the kind that oil and forestry firms must cope with when they are criticized for being environmentally unfriendly. My personal observation is that the nanotech community has tended to be a little cowardly in its response to this kind of thing so far. For example, at another conference that I attended, there was a lot of concern that the movie of Michael Crichton's book *Prey*, which is a Frankenstein-like story in which vicious nanobots ultimately consume their creators, would turn the public against nanotech. As things have turned out it seems that there will be no such movie any time soon, so they needn't have worried.

Nonetheless, the claim that nanotech is harmful in a general and significant way cannot be dismissed, because, in the past, similar protesters with similar arguments have successfully lobbied for a moratorium on biotech research. The parallels are too obvious to be in need of further explanation. So a brief guide to the supposed evils of nanotech seems to be in order.

In practice there are three kinds of arguments that are levied against nanotech. These are discussed below along with the counter arguments.

The Gray Goo Problem This is the silliest argument brought against nanotech, but it is an objection that probably sounds quite sensible to people who have not bothered to educate themselves about the technical side of nanotech. Anyway, it was Eric Drexler who dreamed up the idea of Gray Goo, so maybe the nanotechnologists have only themselves to blame.

What is Gray Goo? Well the idea is that, as we discussed in Chapter 1, in a world in which Drexlerian technology has become a reality, much will be achieved by molecular assemblers that take a basic feedstock and turn it into anything that we desire. Now what happens if these assemblers run amok? In particular, what if they start gobbling up all the matter in a particular geographical area (or in the planet, or in the universe?) And what if they then convert this matter to a primal disorganized slime that is no use to anyone? Then we have Gray Goo, and not much else..

The accidental conversion of the entire universe to Gray Goo is, if it were possible, would be the greatest single cataclysm that one could possibly imagine and it certainly makes a good theme for a science fiction book. (There is even one science fiction book that imagines the end of time in the Christian sense as occurring as the result of the universe being converted to Gray Goo.) But outside of sci-fi, one might reasonably respond to a concern about the Gray Goo problem in much the same way that one might respond to a relative or friend who has a tendency towards fear and paranoia. One should admit that there is a problem, but gently assert that the probability of this problem occurring is so low that it is hardly worth bothering about. Conversely, one should point out, but not taking this small risk, a lot of opportunities would be missed. This will not convince everybody, but there is nothing much that can be done about that, I fear.

Perhaps there is room for some kind of regulation that insists that self-assemblers should not be capable of reducing the world to Gray Goo—something similar to what Isaac Asimov set out in his laws of robotics. Given that self-assemblers do not yet exist, and may not do so for decades, it is not even clear what such a regulation would actually say or how it would be implemented.

In short, Gray Goo is probably not something that anyone should be spending a lot of time worrying about right now. For firms building the kinds of products on which this book focuses it is, and will remain, an issue that could be safely (in every sense of that word) ignored, if it weren't for the fact that Gray Goo is likely to become a propaganda vehicle for the neo-Luddites.

Enough Bill McKibben's book, *Enough*,[50] got much publicity at the time it was released, as did an article with a similar theme in the April 2000 issue of *Wired* magazine by Bill Joy, who was then the CTO of Sun Microsystems. Although McKibben may also worry about Gray Goo for all I know, he mostly objects to nanotech on quite different grounds than did Drexler. For them objections to nanotechnology are more a matter of philosophy than science.

The basic argument runs like this. There are certain things about us that make us human and we mess with those at our peril. Thus it is highly desirable to develop technologies that will let us live a few years longer or prevent us from suffering when we are sick. However, it is highly undesirable to develop technologies that will make us immortal, because in doing so we make ourselves less human, as the essence of humanity is that we are mortal. Indeed, while immortality might be the most dramatic impact that nanotechnology could bring about, potential enhancements to our lifestyle that make us different in some fundamental way from how we are now would also fall under McKibben's and Joy's sword. Enhancing our memory in a dramatic way using nanomemories hooked into the brain with conductive polymers might be something that McGibben and Joy would not like much.

There are a number of criticisms that may be thrown at the *Enough* argument. One possible argument is the same as I have used to critique the Gray Goo problem and that is simply that it is not going to happen. However, this is in practice a weak argument. While immortality is a long way off, memory enhancement of the kind I just mentioned is pretty much here with us right now. Which raises the big question of where exactly should you stop technological progress and what exactly constitutes our humanity. Another question is how we should stop technological progress. The most obvious solutions to this problem seem pretty inhumane in themselves.

These are, in essence, philosophical problems and therefore a discussion of them lies well beyond a book of this kind. They are, however, practical problems for the businessperson, because they can be used to counter the whole notion of nanotechnology in general and nanobusiness in particular and once again could be adopted in the programs of influential parties and politicians. In addition, the *Enough* argument has some genuine appeal. Who hasn't wanted to go back to simpler times? One simply cannot counter this argument with the claim that it is nonsense. One can, however, make the point that those who do not wish to benefit from the more dramatic potential benefits of nanotech need not do so. And at least for me, the implied assumption that in a world transformed by nanotechnology, people would somehow be less (or indeed more) spiritual is a non sequitur.

Nanomaterials Are Especially Harmful Pollutants This is certainly the most reasonable objection to nanotechnology. After all, nobody is going to argue that new materials should just be assumed to be safe. It will easy for detractors from nanomaterials to point out that new materials such as lead additives and asbestos were once thought to be highly useful substances and were later discovered to have potentially fatal consequences.

The objections that the detractors from nanotech in this regard seem to have are twofold. First, if the next big push forward technologically is going to be a materials one, there will be a lot more different kinds of materials and that means more possibilities of things that can go wrong. Second, nanomaterials operate at about the same size level as the molecules in our body, which means that there may be a biological impact of nanotechnology that is intrinsically different than anything that has come before.

It is hard to argue in any fundamental way with either of these two points, since they are essentially the same points that nanotechnology advocates make. There are going to be lots of wonderful new materials and, since nanomaterials are bio-sized, they have the potential to create important new directions for medicine. Does this mean that new government regulations are required? There are plenty already on the books to handle toxic substances and special ones for substances intended to be used in medicine. But from a purely business perspective, don't be surprised if nano-regulation becomes a key area of debate and lobbying in the next five to ten years.

Nanotech: A Future Bubble? There is plenty of reason to get excited about nanotechnology, although not too excited. Enthusiasm is no substitute for analysis, and there are still plenty of unemployed ex-CEOs from dot-com firms to prove this point. I am frequently asked by both my clients and the journalists with whom I speak what the major societal impacts are going to be in the near-term and my answer is usually a disappointment to them. I don't see too many. Instead, as I have already indicated, I see nanotechnology in the short term as being an important enabling technology for the megatrends that I have referred to above.

In the long run, again as I have indicated above, it may well be that nanotechnology becomes an epoch-defining technology in the sense I have defined it above, but one should be quite careful in pronouncing any technology epoch making. Forty years ago, when I was at grammar school, only the most hardened skeptics would have argued against the proposition that by the early 21st century there would be colonies on the moon churning out products and profits from extraterrestrial mining, gravity-free manufacturing, and space tourism. Today this kind of thing looks about as futuristic as it did in the 1950s and 1960s. Nobody, it seems, really thought through all that it would take to make the Space Age dream a reality. At roughly the same time that the coming space age was being touted, we were also hearing a lot about the coming Atomic Age, in which, not only would we all be the beneficiaries of electrical power that was "too cheap to meter," but also of miniature atomic power stations that could be installed in homes, offices, or even cars.

It all sounded as wonderful and as certain to come about as some of the prognostications about nanotechnology today. It is all too easy to get caught up in the excitement of the moment and hence to misunderstand the true costs, technical problems, and real benefits associated with a new technology.

In addition to getting wrong the actual point in time where the epoch-defining technology really takes off, there is the danger of pouring resources into a technology that only seems epoch-making, but is in reality just a "tech bubble." There are plenty of rich men and women around today who got that way by investing in and/or finding jobs in the Internet industry early in its evolution and pulling out before the "dot bomb" fiasco. There are even more who will have difficult retirements, because they thought the party would last forever.

Even where we are discussing a genuine epoch-making technology, it is important for business decisions to draw some conclusions about just how long that epoch is going to last. A long time ago this did not matter very much. For example, ancient Egyptian technology changed very little for 3,000 years, a time period for which few business people now or (presumably) then ever planned. By contrast, the information age seems to have lasted fifty or sixty years, from the first giant computers to the commoditization of the PC and optical networking. In fact the risk of entering a new epoch-making technology market is growing because the "the next big thing" gets replaced by "the next big thing after that" at an increasingly rapid pace. (The famous inventor, Ray Kurzweil, has written an excellent book, which I cite in the appendix, about how this increasing pace is taking us rapidly to a point at which will become, in effect, transhuman.)

I talked with a venture capitalist recently who was skeptical about the prospects for nanotechnology and seemed certain that nanotech was "just another bubble." This is certainly possible, and at the beginning of this book I quoted the famous venture capitalist, Vinod Khosla, as saying that nanotech will someday go through a bubble, much like the dot-com bubbles of yesteryear.

There are few signs of this bubble at the moment. Indeed, as I write these words, VC funding is down for nanotech on a year-on-year basis and there are probably as many skeptical articles being published in the press about nanotech as bullish ones. This is not the way things look as we enter a bubble.

It is true that a few years back there was a fashion of putting "nano" somewhere in the name of a firm, which is certainly the kind of thing that happens during a bubble. I personally carried out consulting work for one firm that changed its name to include the word "nano," although it did not change the product it was selling—or trying to sell—even though that product did not utilize nanotechnology in any way. And the firm, I am sure would failed just as fast had it kept nano out of its name!

This little semantic bubble as it were, aside, I suspect that all this talk about a nanotech bubble is one of those cases where people are fooled into believing that the future will be identical to the past. But unlike the dot-com disaster in which there were hardly any barriers to entry, nobody sets up a nanotech business lightly. They need to buy AFMs and the like and this costs real money. Nonetheless, this is not an absolute guarantee against a future nano-bubble. And perhaps as a sector, nanotechnology may actually need a bubble to push it forward. However, from the perspective of the individual businessperson (and the individual investor), it is vital to take a hard-nosed look at the real size, time frames, long-term growth prospects and business characteristics of the nanotechnology markets that you are attacking. Even in the midst of a bubble, don't be surprised if this kind of analysis doesn't reveal the hard truth that the immediate prospects for your nanoproduct are niche-like.

On the other hand, while there is a near certainty that some nanotech firms will follow the same overoptimistic path to oblivion taken by photonics firms before them, I also sense that others are underestimating just how revolutionary some of this stuff really is and how big it could be in terms of revenues. As a result, caught between the overoptimistic and the overpessimistic, market forecasting of nanotech markets can be frustrating activity. Recently, I was told by one well-informed industry insider that my projections for carbon nanotube field emission displays were preposterously high and by another equally well-informed industry insider that they were unconscionably low. Forecasting of nano-enabled products is a business that doesn't get you much love. It is, however, a necessary one.

Summary: Key Takeaways from This Chapter

This chapter covers a lot of ground, so there are more takeaways than in the other chapters.

1. There seems to be a consensus that nanotechnology is not an industry in the usual sense, but rather an "enabling technology." But given that nanotechnology can be shown in practice to be the amalgam of *many* different technologies, it is unclear what makes nanotechnology a business sector in a unified sense. In the main body of this chapter I analyze a number of different answers to this question and conclude that the vast majority of "nanobusiness" activities falls into three sectors: nanoelectronics, nanoenergy, and bionanotechnology. Each of these sectors has its own strategies and timeframes. Although I agree that commercialized nanotechnology will have a very broad impact throughout the world economy, I believe that most of that impact will come as the result of developments in these three big areas.
2. Much of the remainder of this book will focus on the opportunities resulting form the commercialization of relatively complex nanodevices or nanosystems. I have chosen this route, because I believe that it is in the higher value-added segments that nanotechnology will generate the biggest opportunities. However, there can be little doubt that the "low-hanging fruit" of nanotechnology—the areas where money will be made in the next two to three years—will be in the materials segment, and there are many interesting nanomaterials that are near or at the level of commercialization. In addition, it is interesting to note that the line between nano*materials* is being blurred by the arrival of smart materials, and this may give some materials firms reasons to rethink their strategies.
3. MEMS is often discussed along with nanotech under the heading "small-tech." There may be some justification for this, notably the relationship between NEMS and MEMS. However, otherwise it is somewhat confusing, since MEMS is a different technology to most of nanotech and the MEMS industry is more developed with different business models to those currently being pursued in nanotech.
4. There are four different kinds of nanotechnology opportunity, each associated with its own business models. Accidental nanotech is just that, while evolutionary nanotech covers early products such as nanoengineered cosmetics that enhance existing products. The big

opportunities, ones that will help create the next Intels and Microsofts, are the ones that fall under my definition of revolutionary nanotechnology, and cover genuinely new products. However, such products are not likely to lead to huge changes in society in the way that some boosters of nanotechnology expect. Such changes will come only with what I call disruptive nanotechnology, which is more in the Drexlerian mode, but unlikely to become a reality for well over a decade.

5. Intellectual property has become crucial to many of the early business models for commercial nanotech and patent attorneys are very much to the fore at nanotech conferences. Part of this has been because, at least in the United States, there has been a certain amount of chaos in filing nanotech patents and a growing concern in the nanotech community that there will be many lawsuits in the future as a result of overlapping patents. However, there is also a belief, although sometimes unexpressed, that it may be possible to build a business model for small nanotech companies that relies mainly on licensing IP. This seems hard to justify given the inherent ability of nanotech to reach the same performance characteristics through quite different materials/technology platforms.
6. Many of the countries around the world have national nanotech programs that vary in character from country to country. The goals of these programs are to promote R&D in favored areas, especially areas in which the government traditionally has a role, such as defense. In some instances these programs have more of an "industrial policy" aspect to them; that is, they are targeted towards making the nation concerned a power to be reckoned with in nanotech. There can be little doubt that such programs will help to pump prime nanotech activity to some extent, but industrial policy programs of this kind have, in the past, often failed to meet expectations and may actually have distorted production away from what the market actually needs.
7. Nonetheless, in the United States and elsewhere, government funding is an important source of money for nanotech and various government agencies are actively involved in providing that funding. Angel investors are another important source. Venture capitalists have shown a mixed reaction to nanotechnology so far. A few VCs are actively involved in funding nanotech firms, but most are skeptical
8. Safety issues are emerging as a PR issue for nanotech firms. Some of the issues being raised are, frankly, ridiculous, although this does not

mean that they won't resonate with the public. An example of such an issue is the Grey Goo issue. However, other concerns about nanotech and health are more serious, as for example, the nanopollutant issue. The best way to handle this kind of problem is probably through existing laws and sizeable firms actively involved in the nanotech business may find themselves compelled to lobby to ensure that this is the way that such problems are handled.

9. Although there is much talk about a nanotech bubble, there is really no reason to think that there is one. Many of the people in the industry are very cautious about the prospects for the sector and so are stock analysts. A nanotech boom may come in time, although it is not inevitable. If and when this happens it may mean both good and bad things for nanobusiness.

Further Reading

There are few books on "nanobusiness" that explore this subject in any serious way. Articles that appear in the general business press are mostly just that—way too general. One book that is worth a look is *The Handbook of Nanotechnology: Business, Policy, and Intellectual Property Law*, although it is not quite as advertised, but instead a rather general book on business with a strong nano slant. Much of what it has to say would be relevant to any technology business, although it goes into considerably more depth on legal (including IP), regulatory, and financing issues than this book does. In this book, I have avoided the topic of stock market investment in the nanotech sector, but this is, I am sure, a topic of interest to of the readers. There are actually a couple of books on this topic, but by far the best is Darrell Brookstein's *Nanotech Fortunes: Make Yours in the Boom*, which despite the rather exuberant title gives a fairly sober view of the prospects for the nanotechnology sector, including an interesting analysis of how to look for booms and busts. The book by J. Storrs Hall quoted in the previous chapter may also be a good source for long-term nanoproduct trends.

With regard to the government nanotech initiatives, probably the best way to stay in touch with what they are offering is through the Web sites of the many organizations involved. I have given some of the most important URLs. Not surprisingly, the World Wide Web is also probably the best way to stay in touch with the latest news, with sites mentioned in the appendix at the end of this book being especially useful in this regard.

3

Nanotech in the Semiconductor, Computing, and Communications Industry

> The end is in sight for Moore's Law of continually increasing computer performance.
>
> —*Grady Booch, IBM Fellow*[51]
>
> World in chaos...excellent situation.
>
> —*Chinese wall poster from the 1970s*

Introduction: Nanotech and Moore's Law

The semiconductor industry is different from the other sectors that we discuss in this book. In these other sectors, over the next two to three years, nanotech will have a profound impact on the products and services that are produced, the way they are produced, and the way business is conducted. Once nanotechnology has been at work for a decade or more in these other sectors, they will be transformed completely.

But while these other sectors will be changed by nanotechnology, the semiconductor industry will *become* nanotechnology. Indeed, this is happening already. As these words are being written leading edge "fabs" (the semiconductor industry's term for a factory) are producing chips in large quantities with features under 90 nm. The semiconductor talks in terms of nodes. Thus the industry is said to be in full production at the 90-nm node. Meanwhile the first

65-nm node fabs will come on stream this year and Intel has just announced that its first 45-nm node fab will be in operation in 2007.

So in this sense, the semiconductor industry already has its foot planted firmly in the nanocosm. There can be little doubt that by the middle of the next decade or perhaps somewhat earlier, we will be hitting the 22- and 16-nm nodes, at which point semiconductor technology will simply be a branch of nanotechnology and none of the old semiconductor industry rules will apply in the way that they did in the past. At this point it becomes very hard to build chips using the old materials and production technologies.[52]

The most important of the rules by which the semiconductor industry plays is Moore's Law. As many of the readers of this book will already know, Gordon Moore, who was then the head of research at Fairchild Semiconductor, suggested in a 1965 issue of *Electronics* that the number of transistors that could be crammed onto a chip would double every 12 months or so and that the cost of each new generation of chips would remain roughly the same. This time frame was later changed to 18 months. This relatively innocent sounding statement underlies the extraordinary achievements of the semiconductor industry over the past several decades. Today we have desktop computers that cost more or less the same as a computer from the 1980s, but have many times the power, because of Moore's Law. We have networking technology in our businesses that operates at speeds that would have seemed mere fancy a decade ago, because of Moore's Law.[53] There are many more such examples.

These examples give some sense of how important Moore's Law is to the semiconductor industry and the business of electronics more generally. In fact, Moore's Law is at the core of the entire economics that drives the semiconductor industry, which operates on the assumption that it can it can move to new generations of chips on a regular basis, charge a premium for the latest generation of chips, and then move on to the next generation of chips, as earlier generations become commoditized. As Dr. Moore himself noted, much later in his career when he was chairman emeritus of Intel, the firm he helped found, Moore's Law has become somewhat self-fulfilling and that chipmakers "have to stay on [the] curve to remain competitive, so that they put the effort to make it happen."

When Moore's Law Fails

With so much riding on continuing down the path set by Moore's Law, it is clear that if Moore's Law runs out of steam, the semiconductor industry faces a crisis. And, as it happens, Moore's Law *is* running out of steam. The industry has now scaled down the devices that it makes to the point where it has become increasingly difficult to build the next generations of chips that follow Moore's

Law using conventional tools and materials. It appears to me that the industry is actually scaling faster than one might have expected some time back. The International Technology Roadmap for Semiconductors (ITRS), which is the consensus forecast produced by the industry every year for both production and chip technology, seems to lag what is actually happening in the field, especially if one looks at pilot plants and not just full-scale production.[54]

"What comes after Moore's Law?" and "When will Moore's Law, come to an end?" are questions that have been asked in the semiconductor industry for many years. Indeed, Moore himself envisioned the petering out of Moore's Law. However, until recently these questions were asked primarily by academics, the semiconductor physicists, and professors of engineering. "The End of Moore's Law," made a good *Scientific American* cover from time to time. However, it certainly didn't trouble engineers working in fabs, and wasn't given a thought by practical businessmen, investment bankers, and venture capitalists.

As Moore's Law has pushed chip manufacturing into the nanocosm—that is, as the semiconductor industry has moved past the 90-nm node—the industry has found itself in growing trouble, as the old silicon/CMOS paradigm has increasingly failed to obey Moore's Law. The problems that the industry is experiencing as it tries to push Moore's Law further are fourfold: too much heat, a lack of high-volume manufacturing methods, a materials crisis, and quantum/atomic level statistical fluctuations. These are explained a little more fully in Table 3.1.

Nanotechnology offers a new way forward. One that gets potentially gets round many of the difficulties set out in the Table 3.1. As we shall discuss later in this chapter, there are several new nanoengineered cooling systems that are being devised for chips. The nanotools that we took a look at in Chapter 2 will ultimately scale to high-volume production. And there are new nanomaterials are also appearing that could avoid the materials "zoo" that we find in the semiconductor industry today and which will be more suitable for nanoscale chips in terms of avoiding quantum effects.

Thus nanotechnology could potentially save the semiconductor industry and create a lot of new business opportunities in the process. However, the industry will not give up on the old ways easily, nor should it, given the amazing commercial success it has had with the silicon/Moore's Law approach. Ultimately, nanotechnology will, almost certainly, radically change the way that chips are built and the materials that they are created from. Nobody expects this transition to occur fast, because the core silicon ways of doing things in the semiconductor industry are now thoroughly entrenched. The major semiconductor firms are still investing billions of dollars in the latest fabs that use relatively conventional manufacturing processes and materials. They are not going

Table 3.1
The End of Moore's Law: Bigger Troubles at Smaller Nodes

Problem	Description
Overheating	More transistors on a chip enable faster processors, but faster processors generate more heat and Intel has actually had to abandon a new line of high-speed processors as a result of chip overheating because of this problem.
Chip manufacturing problems	As the feature sizes on chips become smaller they become increasingly hard to create using conventional lithography processes. Most obviously, this occurs when feature sizes are significantly smaller than the wavelength of light. This means that new kinds of lithography need to be developed or some of the nanotools described in Chapter 2 need to be brought into play. Unfortunately, none of the approaches described in Chapter 2 have yet been developed to the point where they can be used to produce chips in volume in the way that the optical lithography used today by the semiconductor industry.
The materials "crisis"	An increasing number of materials must be used in the chip to make it viable at smaller nodes. For example, new materials are being proposed to produce faster interconnects and better insulation between logic gates. Intel has announced that 51 different elements will be used in its 65-nm node chips. This is an extraordinary number when one considers how many elements there are to begin with and how many are entirely unsuitable for any manufacturing process such as semiconductor manufacturing. A decade ago, about 15 elements were used in a chip. At the 65-nm node and below, it seems that chips will be a thing of shreds and patches, raising issues about connectivity between different materials and the complexity of necessary manufacturing processes.
Quantum effects and atomic level statistical fluctuations	We have already met this problem in the general context of nanotechnology, although one of the examples we gave was specifically from the semiconductor industry—the example of soft errors. At the 95-nm node and below, these effects begin to come into play and affect performance. This problem is expected to a major problem by the time the industry reaches the 22-nm node.

to abandon this investment just because self-assembled carbon nanotubes offer some theoretical route to incredibly fast processing. *An important corollary of this important fact is that any nanotechnology solution that isn't designed to work in close harmony with the existing CMOS infrastructure has no chance of commercial success.*[55] In a decade or so there may be enough momentum behind carbon nanotube electronics for an independent CNT electronics sector to emerge. Such a sector could tap into opportunities that the CMOS compatibility requirement would preclude, as for example, sensors and computing devices that are hybrids of organic molecules and carbon nanotubes. These sorts of devices would require an entirely new kind of fab, one that was specifically built with carbon nanotube electronics in mind.

Such a fab will probably not be built for another ten years,[56] even if this means that the world will miss out on some interesting new devices. For the time being, however, in terms of performance and cost, CMOS is hard to beat and the semiconductor industry will not entirely abandon it or conventional ways of making chips for many years to come, nanotech or no nanotech. Unfortunately, this kind of thinking is often taken to extremes and it is by no means unusual to come across people in the semiconductor industry who take the position that there is really no need to spend much time worrying about entirely new ways of doing things.

For some, this may actually be a reasonable stance. For example, there are still plenty of firms around that make discrete semiconductors or EPROMs or simple embedded processors. These simple electronics product will go on selling in mass quantities for years and will do so with no need of an assist from nanotech. Other firms may find that redesigning chips may help avoid the problems of scaling. This has worked with DRAM memory in the past, and Intel's response to thermal problems at the 90-nm node has been to move to dual core processors. However, changes in architecture, while they may be successful design strategies for specific chips, are not a general way of avoiding scaling problems. Redesigned chips may also tax the CMOS paradigm as scaling proceeds to smaller nodes. Some architectural changes are so profound that they are as difficult to accommodate at the manufacturing level as entirely new nanomaterials/production technology platforms. I am thinking specifically of 3-D architectures here.

For the rest of the semiconductor and computer industry, the transforming impact of nanotech will be key. This impact will be felt as the result of a number of nanoelectronics research programs, each based on a common materials/technology platform. In what remains of this chapter, I will explain the main nanoelectronics research programs and what they offer and then conclude the chapter with a description of where the main opportunities are to be found in nanoelectronics. Before beginning on this path, however, it is worth noting that there are two ways of looking at the nanoelectronics market.

One perspective is to start with the various applications that can be nano-enabled; that is, memory, processing, thermal management, and the like. The other is to start with the various technology/materials platforms that make up nanoelectronics; that is, nanotube electronics, spintronics, and molectronics, to name a few. Normally, I would have chosen the applications perspective, since markets and business opportunities are created by the market need for applications, not technology platforms. However, in this case, I am analyzing the market from the technology perspective simply because this is the way that nanoelectronics is actually organized. Firms tend to develop particular nanomaterials/technology platforms with interesting electronic properties and then look at the applications that are best suited to those properties. In other words, there are carbon nanotube electronics firms, but not nanomemory firms. Whether this may change at some time in the future is an interesting question, but not one I plan to take up in this book.

Nanotubes, Nanowires, and Nanoelectronics

Carbon nanotubes (CNTs) are the material that seems to exemplify the high potential of nanotechnology. Nanotubes are effectively tubes made of graphene sheets with carbons in a hexagonal arrangement in that the structure can be imagined as a rolled up sheet of chicken wire. The way the wire is rolled up (the hexagons can spiral up the tube in varying degrees or circle it in two different ways) dictates whether the tube is conducting or semiconducting. Carbon nanotubes come in single-walled (SWNTs) and multiwalled (MWNTs) flavors[57], but it is the SWNTs that are most researched for electronics applications.

While carbon chicken wire may not sound too promising, CNTs actually boast an extraordinary array of interesting properties (see Table 3.2). In the context of electronics properties that are worth mentioning are strength, high ability to serve as a heat sink, and the ability of CNTs to serve as highly efficient conductors, semiconductors, or even superconductors. This last mentioned capability is also something of a challenge. Manufacturers of nanotubes are seeking the most cost effective ways of producing nanotubes that are mostly conductors or mostly semiconductors, since an uncontrolled mix would obviously be of little use to produce standard electronics products.

Table 3.3 summarizes the applications that have been conceived for carbon nanotubes. Important takeaways from this data include the following:

- *"Low-hanging fruit" in sensors.* CNTs are already being used in sensors, where they can serve as sensitive detectors of chemicals, biomolecules, motion, stress, and pressure.
- *CNTs in FPDs?* Meanwhile, flat panel television displays based on carbon nanotube field emission devices should start to appear in stores in

Table 3.2
Properties of Carbon Nanotubes

Property	Comment
Strength	Greater than 50 times the strength of steel
Current density	~ 10^9 A/ cm^2
Elasticity	1 to 1.2 TPa
Thermal conductivity	More than twice that of a diamond
Density	About half that of aluminum
Chemical reactivity	Functionalizes like graphite
Thermal stability	Stable to 2,700 °C
Chirality	Metallic or semiconducting

Source: Nantero.

Table 3.3
Summary of Carbon Nanotube Electronics Opportunities

Product	Market	CNT Advantages	Timeframe	Selected Firms Involved
Logic/ processors	No commercial products at present. Unclear what applications would be served with such products.	High level of conductivity makes for very fast processing. Enables Moore's Law to be pushed beyond the capability of CMOS.	Probably no commercial products for up to a decade. But closer than noncharge-based solutions.	IBM, NEC, and Infineon have done important R&D in this space.
Computer memory	Several approaches to CNT-based memory being developed. Nantero's solution close to being marketed.	CNTs could make for high capacity, nonvolatile memory that could substitute for both SRAM and Flash.	In 2006 for first products at earliest.	Nantero

Table 3.3 (Continued)

Sensors	Could be used in a broad range of sensor applications in energy, homeland security, and medical applications.	Fast, small and (potentially) low-cost sensors.	Currently available.	Nanomix
Displays	Primarily FEDs, but CNTs may be used in some types of flexible display.	FEDs combine all the advantages of plasma, LCD, and CRT displays. CNTs may improve physical qualities of flexible displays.	In 2006 for first FED products.	Samsung
Interconnects	Eventually interconnection of devices on chips.	The need for high-speed interconnection is increasing as processor speeds increase.	In 2010 for first commercial products.	Fujitsu, IBM
Thermal management	Various applications, but especially cell phones, other handhelds, and mobile computers.	CNTs have good thermal conductivity properties and can make excellent heat sinks.	Reportedly carbon nanotubes already being used as heat sinks.	Intel/Zyvex
Packaging	Limited applications for ESD and EMI shielding for device and IC packaging.	Good conductivity at low filler rates.	Near term.	CNT manufacturers already appear to have suitable products.
Plastic electronics	Carbon nanotubes could be used to strengthen substrates or provide connectivity.	Strength and conductivity.	Maybe 2007.	Eikos

the next year or so. These are being developed by Samsung and Motorola, among others, and will combine the flatness of the LCD screen with the visual quality of an old-fashioned CRT. However, this new type of display will have to compete with existing plasma and LCD televisions. Is the world ready for another television technology?

- *You must remember this.* Another relatively near-term application for carbon nanotubes in electronics is in computer memory. This is an application that is associated primarily with the start-up Nantero, although there are other groups working on CNT memories using somewhat different approaches to Nantero. Samsung and the Max Plank Institute in Germany have both worked on this issue. It is widely believed that nonvolatile memories with capacities as high as a terabyte could be created using carbon nanotubes. However, as we discuss below, there are other ways of producing high-capacity nanoengineered memories.
- *Copper and after.* The semiconductor industry has only just settled on copper as the interconnect material of choice, but this material will begin to run out of steam a few nodes down the road. When this happens, CNTs look promising. Placing them as interconnects has already been demonstrated to be compatible with current IC manufacturing processes. In addition, CNTs promise to dwarf anything that is likely to be achieved in the current search for higher-k materials. CNTs are said to conduct electricity 40 times more efficiently than copper. Conducting CNTs can carry much higher currents than copper—up to 1,000 amps per square centimeter. At 100 amps per square centimeter, copper starts to melt. (Electrons travel ballistically in nanotubes, that is, they do not bounce around through the atomic structure generating a lot of heat. Instead they head like a bullet down the middle of the tube, without generating much excess energy. This is the reason why CNTs can handle such high current densities.) Because of this property CNT interconnects would have the advantage of being both smaller. Ballistic transport also means that the tubes do not follow Ohm's Law, and that resistance does not increase with length in the way that it does with normal conductors.
- *Cool it.* Yet another suggestion is to use carbon nanotubes as a way to control the heat problem that, as we have seen, increases as the industry progresses down the path set for it by Moore's Law. The most obvious way that CNTs could serve in this function is as a heat sink, as CNTs conduct heat as well as diamond. However, there are more sophisticated cooling solutions using CNTs. One firm has demonstrated the

ability to make tiny cooling air currents using sparks generated between nanotubes. However, this approach seems more designed as a way of cooling laptops or smartphones than the chips themselves.

CNTs may also serve a useful role in semiconductor manufacturing in the not too distant future to produce arrays of solid state e-beam generators for e-beam lithography. Other potentially important applications for carbon nanotubes are in flexible backplanes for displays and RF generators (a somewhat speculative application.) All of these applications seem certain to see some level of commercialization in the next five years or so.

By contrast, complete processors built from nanotubes would potentially operate at extraordinarily high speeds, but they are unlikely to see commercialization until at least 2015, if then. As always the semiconductor industry will be looking for innovations in materials, architectures, and processing that can push today's paradigms and fabs just a little bit further, rather than adopt an entirely new manufacturing paradigm that entirely eschews silicon. Nonetheless, carbon nanotube processors that have been built in the lab have been quite impressive. For example, semiconducting CNTs have been used to make field-effect transistors (FETs) showing excellent performance compared with standard MOSFETs and work at the University of California at Irvine suggests that nanotube transistors might be able to switch 1,000 times faster than current CMOS transistors.

Although it is possible to imagine a future when carbon nanotubes form the core material for electronics, that future won't even begin to emerge for a decade or more. Building a few impressively performing transistors in the lab is a very different thing from producing such transistors in large volumes at a price that is also competitive with conventional alternatives. As I have already noted, semiconductor firms are not likely to move away from CMOS lightly and no semiconductor firm has any current interest in building specialist CNT electronics fabs. Instead, CNTs will be used in conjunction with CMOS structures for the foreseeable future. For carbon electronics to emerge as a dominant paradigm that could challenge silicon, there will have to be a significant breakthrough in managing large-scale fabrication using CNTs. Such a development is definitely not going to happen in the near future, but should not be ruled out entirely as something that could develop over the longer term. Approaches with potential include templated self-assembly, clever use of electric fields, or even the use of laser tweezers. Infineon has claimed to have developed an approach suitable for mass production based on growing multiple nanotubes in place. Such an approach would be compatible with current manufacturing approaches.

Motorola demonstrated, in late 2003, a highly parallel technique for growing predominantly semiconducting nanotubes in predetermined locations.

For the moment, however, silicon rules. But according to the market analysis done by my firm, NanoMarkets, the entire nanotube electronics market will be worth $6.4 billion by 2010,[58] with most of those revenues coming from memory, sensors, and displays. Because of the remarkable properties that carbon nanotubes display, they have attracted the attention of the R&D departments at some of the most important semiconductor and electronics firms in the world. These include Fujitsu, GE, IBM, Infineon, Intel, Motorola, NEC, and Samsung.

R&D work on carbon nanotubes began in the 1990s and as the "history" provided in Table 3.4 suggests one key area of research, as well as a source of competitive advantage for carbon nanotube electronics firms, is how to place nanotubes on other structures (most typically silicon nanostructures.) Approaches that have been tried include laser tweezers, growing CNTs in situ using some kind of guide, catalytic methods, and deposition methods. However, chemical vapor deposition (CVD) is the commonest technique currently used for manufacturing with CNTs at the present time.

Combinations of these approaches are also being considered and it is becoming clear that manufacturing may become a key market distinguishing feature, because this can easily determine both yields and costs. The commercial viability of CNT-based products is increased significantly if it can be produced with existing production technologies and in existing fabs.

Another source of competitive advantage has been finding ways to sort out conducting CNTs from semiconducting CNTs. As we have already seen, both types of CNTs are amazingly useful. However, if they are all clumped together, their usefulness is considerably reduced, since conducting CNTs are used for certain applications, while semiconducting CNTs are used for quite different applications. In a clump their usefulness is considerably diminished. A big problem with nanotubes at the present time is that there is currently no sure-fire way of producing only semiconducting ones or only conducting ones. Claims *have* been made about processes that will supposedly produce only conducting CNTs, but these have not been corroborated. Separating the two types is not easy either, although progress has been made in this field. IBM, for example, has managed to manufacture batches of CNTs that are only semiconducting by using a high current to vaporize conducting ones. DuPont has used DNA to sort the two types and an electrophoretic approach has been used in Germany. It should be noted, however, that perfect separation of the two kinds of tubes is not required for most near-term applications. While important in the long run, is not necessarily an issue when looking at application as interconnects, and is far from the biggest problem faced when considering building circuits from nanotube transistors.

Table 3.4
Evolution of Nanotube Electronics

Year	Issue Addressed	Achievement
2004	Placement	Arryx uses 200-beam laser tweezers to organize multiple nanotubes into patterns.
	Separation of metal vs. semiconducting	"Metal" CNT transistor made at the University of Illinois at Urbana-Champaign by using an irregular magnetic field to change the tube from conducting to semiconducting.
	General performance	Researchers at the University of California at Irvine demonstrate record switching speeds in a CNT transistor. 2.6 gigahertz was achieved and the researchers believe terahertz frequencies to be possible, 1000 times faster than modern ICs.
	Placement	Researchers at Northwestern develop a technique using combined AC and DC fields for placing multiple CNTs to span an electrode gap.
2003	Contact resistance	Schottky barrier probably overcome by IBM using palladium wires and CNTs, making a device capable of carrying unprecedented currents at a modest voltage.
	Placement	Researchers at the Technion-Israel Institute of Technology build a CNT-FET using DNA to guide placement and also act as a template for metal connectors.
	Separation of metal vs. semiconducting	Dupont uses DNA to sort CNTs according to diameter and electrical properties.
	Placement; contact resistance	NEC claims a process that can reliably make CNT-FETs with 20 times the transconductance of standard MOS transistors. The company predicts commercialization by 2010.
2002	Catalyst contamination	IBM develops a CVD variant using vaporization of silicon carbide for growing CNTs on silicon without a catalyst.

And to help with the future evolution of nanotube electronics, the Institute of Electrical and Electronic Engineers (IEEE) has formed the IEEE Carbon Nanotube Quality Testing Study Group to "define uniform protocols for nanotube characterization to assure nanotube consistency and reliability."[59]

Finally, let's take a look at a couple of supposed objections to CNT-based electronics which, it turns out, are not as serious as they seem.

One of these supposed problems is "doping." While the semiconductor industry depends completely on the properties of *doped* silicon, no known method exists for easily doping CNTs. This may seem like a disadvantage, but much the same effect as doping in silicon can be achieved by applying an electric field. The potential of this approach has been best illustrated by General Electric, which put a double gate under a nanotube, one at each end. Biasing the gates with opposite charge created an excellent diode (or two, in fact: either n-p or p-n, depending on polarity). Biasing both gates the same way created an n- or p-type transistor. This so-called dynamic doping suggests interesting new architectural approaches.

General accounts of the current state-of-the-art in CNT manufacturing often make a big deal about the high cost, limited production, and variability in physical characteristics, such as purity and tube dimensions. While such issues cannot be dismissed entirely, they are much less important in electronics than they are in applications where large amounts of carbon-nanotube-based materials are being deployed, such as in coatings.

It is true that carbon nanotubes are expensive. (Popular accounts of nanotechnology, usually compare the price per ounce of CNTs to the price per ounce of gold.) However, the price for CNTs are coming down rapidly, reflecting the evolution of nanotechnology from a research area to a business. Formerly, CNTs were made in very small batches and sold to limited numbers of research and industrial labs. Manufacturers had to charge large amounts, because manufacturing costs were spread across such small volumes. This is changing as (electronic and nonelectronic) applications for CNTs are being discovered and nanotube manufacturers and economies of scale are beginning to kick in. In any case, most electronics applications use only small amounts of CNTs, so that they may not be especially sensitive to the price of CNTs once they have fallen below a certain level. When I talked with manufacturers who were planning to bring televisions to market using CNT-based field-emission devices they told me that such televisions would use only a few dollars worth of CNTs, although this is in a product that will presumably fetch a few thousand dollars at retail.

Limited production and variability in physical production are also not the problems for nanotube electronics that they are for other applications of nanotubes. This is because in most applications for carbon nanotubes in electronics and semiconductors, the nanotubes used are grown specifically for the purpose, so there is considerably more control over physical characteristics and the availability of supply of CNTs from third-party sources does not apply. In any case, as the opportunities for nanotubes both within electronics and in other applications become increasingly evident, carbon nanotubes are becoming easier and easier to obtain in significant quantities. This is no longer a market where materials are supply constrained.

Notes on Nanowires

Nanowires are often grouped with nanotubes, more for reasons of geometry than anything else. At this level, nanowires and nanotubes do have a resemblance. They both share the fact that they are essentially single-dimensional structures, with quantum effects coming in to play in their performance and capabilities because two of their dimensions are nano-scale. Not surprisingly given all this, nanotubes and nanowires also share some production issues, such as how best to create them and how to connect them up in a scalable manner.

Beyond this, nanowires represent a different paradigm for nanoelectronics to nanotubes, one that competes for more or less the same applications as nanotubes. These applications include sensors, interconnects, memory devices, and (potentially) logic devices and processors. And despite the geometrical similarities, nanowires are really quite different structures to nanotubes and have significantly different properties. Perhaps the biggest difference between the nanowire research program and the nanotube research program lies in materials. Nanotubes *can* be constructed using a number of materials and university labs have come up with interesting list of materials from which nanotubes can be constructed. However, nanotubes are usually made of carbon and from a commercial point of view, they almost invariably are. Nanowires have been grown out of metals, traditional semiconductors such as silicon and gallium, and a variety of polymers.

Nanowires are also quite diverse in the ways that they can be created. They can be constructed using a variety of high-resolution lithographic techniques, they can be grown with CVD (like CNTs) or they can be built up through the self-assembly of appropriately chosen units. It is this latter approach that has yielded some of the most intriguing results. Unlike nanotubes, which might be ordered through self-assembly but not created that way, periodic nanowire patterns with wire and interwire widths of just a few nanometers can be made with self-assembly. Other approaches to manufacturing nanowire structures include nanoimprint lithography (NIL) and molecular beam epitaxy, the latter being a process familiar to the semiconductor industry and capable of creating very regular arrays of nanowires. NIL is already being used to manufacture optical components and, as we noted in a previous chapter, has attracted much attention as an up and coming manufacturing technology for devices with nanoscaled features.

While no one firm stands out as *the* nanotube company, although several companies, such as IBM and NEC, have made impressive contributions to the commercialization of nanotubes, one firm does stand out in the nanowire business. This firm is Nanosys, and its technology is based on ideas developed by Professor Charles Leiber at Harvard University. Nanosys has also assumed an important role in the history of nanotechnology commercialization, because it was thought by many in the financial community that when Nanosys did its initial public offering (IPO) and became a public company, this would mark the beginning of a nanotech boom, in the way that Netscape's extraordinarily successful IPO marked the beginning of the Internet boom. Unfortunately, this was not to be. Nanosys withdrew its IPO at the last moment.

Returning to the matter at hand—the nanowire market—Nanosys is by no means the only other firm pushing nanowire electronics. Others include NEC and Hewlett-Packard. Whether Nanosys' prominence is maintained as the nanoelectronics market begins to take off remains to be seen. However, its current activities and collaborations are a good indication of where nanowires will be used and how they will be used. Nanosys has been involved in various collaborations that use its nanowire technology in photovoltaics, sensors, and computer memory. Moving forward in time nanowire transistors may prove easier to build than those built from nanotubes. Nanowires also show a lot of promise for interconnects. Silicon nanowires coated with nickel have been built at Harvard with this application in mind and have proved to have very high conductivity. Nonetheless, there is little doubt that, whatever the virtues of nanowires may be, it is carbon nanotube electronics that has a lot more buzz in terms of funding and commercialization.

Spin and Nanoelectronics

Roughly speaking, spintronics uses the quantum mechanical property of spin in much the same way that regular electronics uses charge. (Spin is strongly associated with the more familiar property of magnetism.) Exploiting control of electron spin has been a favorite of researchers for many years and almost every major electronics company has played with the idea at some time or the other. The one established product that might be considered the child of spintronics is the giant magnetoresistance (GMR) read/write heads already used widely in disk drives. Magnetic RAM (MRAM) is just becoming commercialized. This new type of low-power-consuming, nonvolatile "nanomemory" is being produced or being developed by some of the biggest names in the semiconductor industry. These include Freescale, Hewlett-Packard, Honeywell, IBM, Infineon, NEC, Sony, and Toshiba.[60] GMR and MRAM are inherently nanotechnological because of the thin layers required in building devices.

For logic and processors, spintronics remains is still in its early stages. However, some of the most important semiconductor firms are also extremely interested in taking spintronics to the next level. IBM, has announced that it is working with Stanford University to create a Spintronic Science and

Applications Center. Intel has been cooperating with the University of California at Santa Barbara, which is considered by many to be at the forefront of spintronics research. The (U.S.) Semiconductor Industry Association has been lobbying for a national research institute for nanotechnology and has listed spintronics as one area that such an institute would concentrate on. DARPA, for instance, is providing $31 million for spintronics research grants. The objective of this research will be to build fast devices called resonant-tunneling diodes. In addition to universities in the United States, universities in Germany and Japan are doing significant work in this field.

Spintronics appears to hold out the prospect of very fast switching and few if any thermal or power consumption problems. But logic and processors make extra demands on spintronics, beyond what is required for memory. Specifically, we need a design for a spin transistor. Some experimental work has been done in this area with spin valves. These are based on hindering, to a greater or lesser degree, the progress of electrons based upon their spin. (Spin valves used to be a popular candidate for MRAM but now seem to have fallen largely out of favor in that application, with tunnel junctions being more popular.) To date, however, spin valves do not show promising performance.

To commercialize spintronics in a wide variety of product types what will be needed is cost effective materials and manufacturing platforms. Today the consensus is that the best available materials for spintronics switches/transistors are semiconductors that are doped with small amounts of ferromagnetic metals, such asmanganese, chromium, iron, or cobalt. But creating spintronics semiconductors has proved difficult and new machines are being designed for spintronics manufacturing. A few firms are already offering commercial manufacturing systems for MRAM. It is possible that the first commercial use of spin in logic will be in quantum computing, where superposition of spin states can be exploited.

NanoMarkets estimates that the market for spintronics products will reach $9.1 billion by 2010. Most of those revenues are expected to come from the computer memory sector.

Nanoelectronics with Molecules

The nanoelectronics platforms that we have discussed so far, although quite revolutionary, bear a striking resemblance to the good old-fashioned silicon microelectronics, in that exploit the electrical and magnetic properties if fairly simple materials. All of the approaches we have discussed to date have also reached a point in their commercialization, where you can be reasonably certain that something will come of them. Molectronics (sometimes called "moletronics") is different, because it has a lot further to go commercially. In fact, it isn't even all that well-defined. Indeed, molectronics is basically a catchall term for electronics that uses complex (often biological) molecules as the main materials platform. There is a considerable amount of interesting R&D in this field that is being performed in both academic and industrial laboratories that involves quantum computing or chemical computing, although this kind of molectronics is so far off in terms of commercialization, that I can so no real business opportunities emerging from this work in this decade. Instead, this kind of thing will remain in university labs or in the labs of the very largest electronics and semiconductor firms for many years to come.

There is another aspect to molectronics, though, one that is closer to commercialization, and which I believe will present some genuine opportunities in the next five to ten years. This approach is one in which switching is based upon the change in state of individual molecules. The change can be based on conductivity as a result of an applied field (much like a classical FET), a conformational change resulting in a change of conductivity (including a complete contact break) or optical properties. In practice, devices may contain many such molecules in an individual switching unit. Although some approaches to this kind of molectronics involve radical changes in circuit design, others are more in line with conventional architectures.

All this diversity—no settled materials, designs, or architectures—is symptomatic of the early stage of development at which molectronics currently exists. Probably the closest product to commercialization that might be classified as molectronics is a memory chip from a firm called Zettacore, which uses complex molecules as material platform, but otherwise doesn't diverge too much from standard chip designs. However, when looking further into the future, when processors and logic may be built using molectronics, molectronics will probably require significant changes in approach to design largely as a result of having to build in a fairly high level of fault tolerance.

One other way in which molectronics will mean a major break with the past is in the area of production. Indeed, it is the hope that self-assembly could be used to produce very low-cost circuitry based on a molectronics that is the key driving force behind work in this field. Molectronics is an entirely different paradigm for electronics from silicon/CMOS and cannot be similarly based *primarily* on lithography. Some direct-write methods could play a role in placing molecules in various nanostructures. In the long run, the most likely approach for high volume of building molectronic circuitry would probably be self-assembly. For obvious reasons, self-assembly is theway to go where biological molecules are involved and in a world in which the molectronic paradigm holds a significant place of honor, opportunities for innovation would be available to those materials (or even biotechnology) firms who can develop self-assembling molecules that also exhibit multiple self-assembling states.

There is something vaguely Drexlerian about all of this, but as we have pointed out, self-assembly does not necessarily mean anything more revolutionary than growing crystals. Thus, although the road will not be an entirely easy one, it would not be the greatest surprise to find self-assembling biocircuoitry playing some role in electronics in a few years time. Obviously, complex molectronic circuitry will challenge capabilities in self-assembly, but hybrid architectures, using some form of template for self-assembly, should provide a stepping-stone. Both lithography and direct-write methods may be used in constructing the template. Hierarchical self-assembly, whereby molecules are designed to self-assemble into structure which in turn self-assemble into others, and so on, also offers great potential and is an area of endeavor as yet little explored. Whatever the methods used a major challenge is likely to be achieving good electrical contacts between the molecules and their connectors. This may prove a challenge and a source of competitive advantage for those firms that rise to meet that challenge.

A look at some recent research developments in Table 3.5 shows some of the promise of molectronics and indicates some of the major issues that surround it today look quite surmountable. It also indicates once again, just how many paths can be taken in terms of materials, architectures, and production modes.

Just One Word, "Plastics"

"Plastic" electronics is based on thin-film transistors (TFTs) fabricated using organic polymer films and it offers some radically new directions for electronics, including the creation of a range of entirely new products that could not be manufactured using conventional CMOS approaches. It also lends itself to exciting new manufacturing processes, notably printing. The field of plastic electronics follows from the discovery that certain polymers can be conductive. This has been know for several decades, but the discovery that polymers could be made much more conductive using dopants is what is driving this area—it also won the team of researchers that discovered it, the Nobel Prize.[61] Using dopants, specific polymers can be tuned between fully conductive through the semiconductive to the dielectric.

Not all printable electronics involves polymers, however. Metallic circuitry can also be printed using nanometallic inks. Metallic inks have been around in the graphics printing business since almost the beginning of commercial printing and metallic inks for printing circuitry are not entirely novel. The big advantage

Table 3.5
Evolution of Molectronics

Year	Achievement
2004	Researchers from the University of Southern California and NASA build a prototype molecular memory device that stores three bits in each memory cell. Each cell consists of a field-effect transistor made from a 10-nm-diameter indium oxide wire. Current applied to a gate electrode produces an electric field around the nanowire, which lowers the nanowire's electrical resistance, allowing current to flow through the nanowire. Molecules of an organic compound adjust the nanowire's electrical conductance to eight discrete levels. Practical use might be possible in five to ten years.
2003	ZettaCore-funded work by ZettaCore founders demonstrates stability and durability in porphyrin molecules that can be used in molecular memories (see main text).
	Infineon claims a class of organic molecules suitable for nonvolatile, high-density memory, with potentially easy production of multilayered memory on CMOS chips. The materials are compatible with aluminum or copper conductors.
	Japanese industry/academic alliance to develop protein-based memory with similar density to best silicon-based devices (8 Gb / cm2) but 1/100 the power requirements. Firms involved include Matsushita, Olympus, and more.
	Significant improvements in connections between electrodes and a molecular memory monolayer are achieved by a joint team from the University of Southern California, NASA Ames Research Center and Rice University. The rather laborious method used electron beams, coated nanotubes, and scanning probe microscopes.
	Researchers from the universities at Lecce and Bologna in Italy make a transistor that operates at room temperature using self-assembly of a derivative a DNA base. Maximum gain was good for a molecular device but still low compared with silicon.
	A molectronics favorite, phenylene-ethynylene oligomers, are shown to be a pretender by researchers at Arizona State University and Motorola Labs. The switchable conductivity previously observed appears to be a result of changes in contact resistance with a gold surface.
	Researchers at the University of Basle, IBM Zurich, and the CEMES-CNRS Lab in Toulouse create the lowest energy single-molecule switch, requiring 10,000 times less power to switch than that needed in transistors currently used in high-speed computers. A porphyrin molecule was used but the switch was mechanical rather than electrical.
2002	Hewlett-Packard creates a prototype 64K RAM measuring one square micron using nanoimprint lithography to create nanowires, which are then incorporated into the crossbar architecture, using rotaxanes as the molecular switch.

that nanotechnology brings to printing circuitry with metallic inks is that inks based on nanoparticles, and today they are almost always *silver* nanoparticles, can be cured after printing at much lower temperatures than other inks. This is a consequence of high surface area of a nanoparticle relative to its volume.[62]

It could be argued that apart from the single issue of the use of nanoparticle inks, a discussion of plastic and printable electronics[63] has no business being in a book about nanotechnology. For example, one of the areas that has generated the most interest in this space is printing circuits using ink-jet printers and conductive organic polymer inks. This approach seems capable of producing very low-cost circuitry in moderate volumes, as long as you are not looking for devices that are especially complex, or *especially small.* Nonetheless, I will discuss printable/plastic electronics in this book for a number of reasons. One reason is that in its main objective to utilize the properties of novel materials to create circuitry, it is very like nanoelectronics. Another reason for thinking of plastic electronics as something that should be discussed in these pages is that it may ultimately evolve into a technology that could genuinely be called nanotechnology. Polymer electronics is, after all, is not all that different from molecular electronics and it is possible that the two areas might ultimately merge, especially as it becomes possible to print nanoscale features. However, even at the current level of development, it is clear that there is something special about plastic electronics in the breadth and depth of the business opportunities that it presents:

- *Novel materials properties leading to new features.* The flexibility of thin-films and plastics more generally suggest new product directions, including roll-up displays, very low cost RFIDs, and flexible sensor, and photovoltaic arrays. Plastic electronics also generates very little heat and uses small amounts of power, alleviating major problems that dog conventional electronics. The approaches to plastic electronics with the greatest commercial prospects are the ones that emphasize these advantages to the fullest. This is because plastic electronics cannot match the performance of CMOS circuits, so its competitive advantage must be found in features and capabilities that cannot be matched by CMOS. There are opportunities to create new firms and new revenue streams for older firms from these products.
- *Disposable electronics.* One of the main new product opportunities that seem to be suggested by plastic electronics is disposable electronics for RFID tags, smartcards, to name a few. Memories using low-capacity magnetic devices are already being added to greetings cards and tickets,

but plastic electronics has the capability to lead to a new breed of smart tickets, greetings cards, tickets, and other products that may include more sophisticated circuitry and even small displays.

- *Electronic paper.* Another important new product direction for plastic electronics is electronic paper. This is a "cool" name for a special kind of display that emulates (literally) the look and feel of real paper. It is thin, flexible, high resolution, and (in some versions) will even feel like paper. The difference is that "e-paper" is still a display, which means that what is being shown on the screen can always be updated electronically, including over a network. E-paper will find applications in a number of areas. There are already electronic book readers that use "e-paper," and e-paper systems can also be used for easily updatable signage in stores and hotels.
- *OLEDs.* Light-emitting diodes (LEDs) have come a long way from the little red and green flashy things that you find on modems and car dashboards. One direction that they have taken is towards high brightness LEDs (HB-LEDs), based on exotic III-V semiconductors (notably gallium nitride). These are already widely in used in home and industrial lighting, flashlights, car headlights, and so on. Another type of LED is currently being commercialized. These are based on plastics/polymers or on smaller molecules and could be the basis of a wide variety of displays, including e-paper. Such organic LEDs (OLEDs) may also ultimately find their way into lighting systems.
- *New processes mean low-cost fabrication.* Conventional semiconductor fabs now cost in the billions of dollars and are expected to escalate for the foreseeable future. The manufacturing processes that will ultimately be adopted for building plastic electronics products remain in flux. However, it is clear that the economics of plastic electronics will be much more attractive than state-of-the-art CMOS manufacturing. Plastic electronics is creating a new paradigm, in which electronic circuits are created either using some kind of deposition technique or printed using ink-jet technologies,[64] stamping, or some process. This means that plastic electronics products could be produced economically in relatively short runs and even customized to the needs of low volume customers. This ability to customize is associated especially with ink-jet printing and other "maskless" printing technologies that do not require huge fixed costs per run.
- *New manufacturing models.* The "printable electronics" concept seems likely to generate some new business ideas. It is possible to imagine,

> although sometime in the distant future, a store, much like today's photocopying store, where circuit designers and entrepreneurs could go to bring their design into plastic realities for test, sampling, or other low-volume requirements. Even if high-speed plastic electronics processing were not available, an engineer or businessperson may want to try out a CMOS concept in plastic in order to get a low-cost answer to certain questions that arise about functionality. This kind of vision may be a little futuristic, but it is possible to imagine plastic electronics manufacturing being used to generate application specific integrated circuits (ASICs) in the relatively near future.

Although plastic electronics cannot yet produce fast transistors and so does not compete in many areas with the regular semiconductor industry, one consequence of plastic electronics' dramatic differences from conventional electronics is that plastic electronics has the ability to change the structure of the electronics industry:

- *Changing industry practices.* If electronic circuitry can be created with something akin to an ink-jet printer, semiconductor firms can start producing their own circuitry at relatively low capital costs. This could reverse the trend towards fabless firms—firms who outsource their manufacturing to foundries, because they simply cannot afford to get into the manufacturing of chips, although the performance limitations of plastic and printable electronics, will mean that the impact will be quite small at first. As I have already noted a more immediate role for plastic electronics in the semiconductor sector will be to build prototypes of circuitry, before recreating it in CMOS. This could radically reduce the front-end costs of designing and building a new chip and may revive the dying art of building application specific circuits (ASICs).[65] The ASIC business may also get a boost in the future from combining standard CMOS and plastic electronics. Raw ICs can be embedded on a substrate and circuitry can then be printed to complete the functionality. This approach has been suggested as a way to reduce the cost of packaging that currently accounts for as much as 30 percent of standard ICs.
- *Changing industry boundaries.* Plastic electronics is about printing with organic or nanometallic inks onto flexible substrates. It therefore opens up the business to firms such as Xerox and Hewlett -Packard, who have long histories in the nonimpact printing business. Some materials firms are also likely to see plastic electronics as an emerging opportunity and

firms such as Dow Corning, DuPont, and the European chemical firm Merck (not the same company as the U.S. pharmaceutical firm of the same name.) There is also a lot of talk about the printing industry getting into the plastic electronics business. This makes sense up to a point: some plastic electronics is being printed using conventional printing technology, so we may well see some printing firms move in this direction. However, most printing firms, especially the smaller ones, will lack both the technical and marketing knowledge of electronics to make a go of this. Although a rather "cool" idea, I am also skeptical that Kinkos is going to go into the print-your-own circuit business, except perhaps in its Silicon Valley locations.

All this may suggest that we are on the verge of finding something new under the sun in electronics. But the truth is that quite a few issues remain to be dealt with in the plastic electronics business:

- *Materials and production uncertainties.* Conductive polymers and flexible substrates that are capable of supporting plastic electronics are commercially available. There is a growing amount of research in this space and no one yet knows what will be the standard materials platforms that will support the plastic electronics of the future. Each material currently being used has its pluses and minuses. Although generally considered to be chemically and thermally stable, exact lifetimes for these materials have most still to be determined and is something of an issue at the present time. How long will they be able to withstand the effects of prolonged exposure to water, air, light, and heat? Similarly while pundits claim that plastic electronics will also be printable electronics, no one is quite sure about exactly what standard forms of printing will emerge. NanoMarkets' research has, for example, revealed serious differences of opinion with regard to the level of development of ink-jet printing.
- *Market uncertainties.* The products that plastic and printable electronics can enable are genuinely novel, and often, so are the markets that they are chasing. Products that can be created using plastic electronics, but could not be created using CMOS appear to have a ready market. But this has yet to be proved. It seems obvious that a large roll-up display that can help turn a cell phone into a computer or entertainment device would ultimately find significant demand. It may also seem obvious that if plastic electronics can get the price of an RFID tag down to one cent, RFIDs will seriously challenge the market for barcodes. Solar

panel laminates seem as if they could improve energy efficiency for buildings, but will they be accepted by builders? But sometimes what is "obviously" going to happen, doesn't happen and it is not clear what can be done about it.

- *The limits of plastic electronics.* While the current generation of plastic electronics is pitched at markets in which it will not compete with CMOS, no one really knows how and where the two materials/technology platforms will ultimately compete. Plastic electronics still seems to be a long way from providing an alternative to CMOS-based processing and logic. On the other hand, some of the theoretical work that has been done suggests that organic material could be used to create processors up to 1 THz, of course, only if one could find the right material.

Table 3.6 summarizes the products in which plastic and printable electronics is (or will be) used. Two important takeaways from this exhibit are (1) that plastic electronics will have an impact on a very broad range of industry

Table 3.6
Plastic and Printable Electronics Roadmap

Product	Comments
Displays for mobile phone and other portable devices	An electronics book reader with a flexible display has been made available in Japan on a limited basis—in Japan people currently read books on PDAs. Subdisplays for mobile phones and small displays for MP3 players already use OLEDs in a big way. But the main thrust of market push into mobile phone and notebook computer displays is likely to occur in 2007 and beyond. Technology will have to improve to enable the pixel density required by main displays on small mobile devices including both cell phones and notebooks.
Advertising displays and signage	There are already nonflexible version of electronic paper for signage and pricing displays. These can be updated over a network. Large-scale OLED advertising displays lie further in the future with about the same time frame as medium-to-large television displays. Outside displays will take more work on materials and encapsulation to make them more durable in exposed environments.
Small consumer products displays	Small plastic displays are already being used in cameras and on automotive dashboards. These are likely to rapidly penetrate the market from here on out.

Table 3.6 (Continued)

Display backplanes, etc.	Display backplanes are not likely to be created using plastic electronics, but rather with nanometallic inks. Both backplanes and color filters for displays may prove cheaper to print than using current technology and many firms are dabbling in this area, with significant revenues expected to be generated by 2007 or even earlier. These backplanes and color filters would be used for existing conventional flat-panel displays, so the addressable market for these products is potentially huge. Some firms are now building very large ink-jet printers to serve this market.
Desktop and other computer displays	Not the main market aimed at by most plastic electronics firms, but there are applications and products that may emerge in a few years as the result of displays being developed for the notebook and cell phone sector.
Solar panels	Despite their obviously very different function, plastic solar panels use fairly similar technology to plastic displays, but reversed. (Light produces energy rather than the other way round.) A few early products will appear in 2006 and will most likely first be targeted towards powering mobile and portable electronics. Building products will be targeted later, although expect some resistance in this area, because plastic solar is such a new technology and the construction industry is understandably conservative about using new technology . Nonetheless, plastic photovoltaic arrays do seem able to reduce the initial cost of deploying solar panels and enable them to be used on products where they could not be used before.
Toys and greetings cards	At the time of writing several firms were developing greetings cards that used plastic circuitry, to enhance the existing trend for intelligent cards—a few such products were available in limited quantities. Toys are not often mentioned as an application for plastic electronics. However, it seems that this area is fairly certain to benefit. Novelty products have often been a starting point in the past for new technologies. This is how both fiber optics and HB-LEDs started out.
RFID tags	A small proportion of the antennas for RFIDs are already printed. However, the long-term objective here is to print complete RFID tags that are sufficiently inexpensive to use on disposable products as a good economic substitute for barcodes. This is probably not going to happen for several years, since it will be some time before printable electronics is developed to a point where tags can be printed in very high volumes. However, by 2008, plastic electronics and/or printed RFIDs using nanometallic inks may have reduced the cost of tags to a point where they are cost competitive with conventional silicon tags.
Sensors	The thrust of technological innovation in nanosensors has not been on using either polymers or printing technology. There are currently some research programs at universities and in industrial labs. However, inexpensive sensors arrays could be the result of this work and would be a major boon in pervasive computing environments and for homeland security applications. As a commercial technology, significant revenues will probably not be achieved until after 2008 or so.

Table 3.6 (Continued)

Computer memory	Polymer-based memories are one of a number of competing nanomemory technologies, but some of the commercialization attempts in this field appear to have been less than successful. Nonetheless, "plastic memory" could eventually be quite important for RFIDs and other disposables as well as in some pervasive computing applications. There are current discussions of this application in the literature, but it is hard to see significant revenues develop from polymer/printed memories until 2009 or after.
Television displays	Plastic or printable electronics seems unlikely to make inroads into this sector for some time and it is, in any case, unclear that the world needs yet another kind of flat panel display. However, some tiny television sets—with primarily novelty value—using plastic screens may appear on the market earlier than 2008. Televisions do have the advantage that pixels can be relatively large compared with computers and (especially) cell phones.
Plastic/printable logic/processors	The exact date for when these will become available will depend to a large degree on what one means by plastic logic/processors. The display backplanes currently being developed for near-term commercialization already involve printable transistors, but freestanding plastic processors that somehow resemble a CMOS processor in power lie a long way off in the future. There does not appear to be any good theoretical reason why polymer processors should be slow, so there may be important, yet to be guessed at, developments in this sector. And polymer electronics ultimately becomes molectronics at a certain stage of advanced development.
Lighting	Flexible OLED lighting displays are more at the idea stage than product stage at present. As with solar cells used in building applications there will be questions about reliability, especially for outdoor use, and there will be an element of fashion contributing to the success or failure of such products.

sectors, and (2) that a significant amount of that impact will be felt in the next two to five years.

On Quantum Dots

I will review the electronic applications for quantum dots.[66] A quantum dot is a structure that is sufficiently small in all directions that electrons contained on it have no freedom to move in a classical sense and are forced to exhibit quantum characteristics, occupying discrete energy states just as they would in an atom. Indeed, quantum dots have sometimes been referred to as artificial atoms.

Quantum dots are the subject of intense discussion by researchers and have actually been used to create commercial lasers. (In quantum dot lasers, the frequency of emitted light can be controlled by changing the dimensions of the dots.) However, quantum dot lasers have not done well in the marketplace. The reason seems to be that quantum dot lasers' main claim to commercial fame was the fact that they were small cheap and didn't need to be cooled, but that, as things turned out, there were easier ways of providing these characteristics in lasers. Despite this initial disappointment, quantum dots may have an interesting future in electronics and may ultimately for the basis of future processors, logic, and memory.

As described Chapter 5 on medical aspects of nanotech, quantum dots appear to have some interesting opportunities in medical imaging and diagnosis. They may also have a role to play in building more efficient photovoltaic cells and in quantum. Other future applications of quantum dots appear to offer genuinely novel solutions for customers, but will be very challenging to commercialize. These applications include quantum computing systems and single-electron (or few-electron) transistors, although the reader should note in the semiconductor device sector, a prototype charge-based memory has been made out of quantum dots by Motorola. Notably, the production process for this memory is not hugely different, and in fact, somewhat simpler, than for traditional floating-gate flash.

Although both commercial applications for quantum dot devices and practical ways of making them in large quantities seem to lie some way off, they are a subject of intense research. Like nanowires, quantum dots can be made using a number of materials platforms and there is no clear indication of what materials could be used that will speed their commercialization. Various approaches have been tried.

Some quantum dots have been fixed on a substrate or have been loose floating using a sol gel approach. The sol gel method mainly has potential electronic applications only when combined, for example, with plastic electronics, where their light-emitting properties can be useful (prototype optical devices have been created by embedded quantum dots in polymers). However, semiconductor quantum dots fixed on a substrate, created lithographically with etching or grown epitaxially (mismatching lattice constants can be leveraged to create quantum dots) have the greatest relevance to electronics.

Nanophotonics

Although this chapter is nominally about electronics, electronics and photonics often overlap. We have already seen this as our discussions of electronics have led us into discussions of OLEDs and lasers. In fact, the impact of nanotechnology on photonics applications is likely to be quite extensive and very much in tune with the normal development of photonic devices, whose evolution in prenanotech days have often been based around new materials developments. Thus, lithium niobate (LiN) helped improved optical modulators and amplifiers and indium phosphide (InP) offered the promise of integrated optics and electronics.[67]

Now it may be the turn of nanophotonics to add its contribution. A number of start-ups have opened their doors in this space and the European Union has just launched a large R&D program focused on exactly this area. The impact of nanophotonics is fairly diverse. NanOpto makes a range of nanoengineered polarizers, splitters, and "waveplates," using a nanoimprint lithography. Toshiba has claimed a breakthrough when it recently announced quantum dot light sources that could transmit single photons and this type of device would most probably be used in quantum encryption. It seems more likely that the first big business for nanoengineered lasers will come from chip interconnection. Until recently, the speed bottleneck in computing and telecom was the speed of the processors. In the last few years, however, the processor speed has reached a point at which it is the interconnections that are now the limiting factor.

In response, semiconductor manufacturers have moved from aluminum interconnects to copper interconnects and are now experimenting with optical interconnection as well as exotic lower-k materials and carbon nanotubes. Optical interconnection could supply more than enough bandwidth to suck up and supply data to even the fastest processors. The requirements for lasers to support such an application would very demanding in terms of size and cost, but the market size is potentially huge. For on-chip applications, the lasers would have to be embedded and their value would be subsumed by that of the entire chip. But an examination of the on-board market suggests that addressable markets for nano-engineered interconnects could be huge. Consider a board with 10 devices on it. If these devices were fully interconnected, 90 lasers would be required. Given that hundreds of millions of boards like this are sold every year, we are talking about a lot of lasers here.

Summary: Key Takeaways from This Chapter

Summarizing, the semiconductor and electronics industry seem to be where complex nano-enabled products will first create large new revenue opportunities. The main things to remember from this chapter are:

1. The semiconductor industry has followed Moore's Law since its creation. Indeed, Moore's Law is fundamental to the industry's economics. However, the industry has now reached a point where making features on chips ever smaller is stretching traditional manufacturing processes and materials to the limits. These issues seem likely to be alleviated as the result of developments in nanoelecronics.
2. A number of nanoelectronics research programs have begun to take on serious commercial importance. These include spintronics, carbon nanotube, and nanowire electronics, plastic electronics, molectronics, and quantum dots. While any and all of these could ultimately take the semiconductor industry beyond the age of silicon, at the present time they all have to be CMOS-compatible. There is no money for building a purely nanoelectronics manufacturing infrastructure at the present time.
3. Several nanoelectronics products are already on the market or soon will be. These include MRAM, CNT-based displays, CNT-based computer memory, organic polymer displays, and quantum dot lasers. Other products seem likely to become commercialized in the next five years. In the traditional semiconductor industry, the sector most likely to be penetrated by nanoelectronics is the memory segment. However, processors that entirely abandon the CMOS paradigm lie quite a long way into the future.

Further Reading

The following books and articles will provide more gloss on nanoelectronics. There are no "basic" books on nanoelectronics, although there is some (fairly superficial) discussion of nanoelectronics in virtually all of the basic books on nanotechnology that were listed at the end of Chapter 1, and *Scientific American's* book of readings on supercomputing[68] provides a lot of useful related material on Moore's Law and scaling processors that will be accessible to the casual reader.

But some specialist books are also beginning to appear. Of the latter, I have found that two are particularly worth studying.

One of these is *Future Trends in Microelectronics: The Nano Millenium*, edited by Luryi, Xu, and Zaslavsky.[69] Despite the title, this book doesn't particularly stress nanotechnology, as a glance at the index will make apparently clear—the term "nanotechnology" fails to appear at all, while the term "nanoelectronics" appears only once. Despite this, there are few other books that give as good an account of the current materials/technology platforms that are now increasingly being taken seriously by the semiconductor industry. Chapters include "The Future with Silicon," "The Future Beyond Silicon," and "Optical and Other Paradigms."

The other book recommendation is *Nanoelectronics and Nanosystems: From Transistors to Molecular and Quantum Devices*, by K. Goser, P. Glosekotter, and J. Dienstuhl.[70] This is a very technical book, fairly far removed from the marketplace in its main focus. However, there a are few other books that discuss in such great deal the engineering considerations for electronics in an era in which Moore's Law in the classical CMOS sense is going into decline.

The reader should consider reading *The Quantum Dot: A Journey into the Future of Microelectronics*, by Richard Turton.[71] This book is now a decade old and doesn't actually talk that much about quantum dots. However, what it does do is provide a highly readable account of all the physics of semiconductors that anyone interested in understanding how nanoelectronics will make an impact is going to need to know.

Finally, my firm NanoMarkets provides ongoing coverage of emerging nanoelectronics markets. For details of our current publications, see http://www.nanomarkets.net.

4

Nanotech and Energy

Nanotechnology could make energy supply lean, green and mean.
—*headline of article on the azonano.com Web site*

Introduction: The Real Energy Crisis

The semiconductor industry may have a crisis pending in its inability to carry the CMOS paradigm forward down the path set for it by Moore's Law. The energy industry has a crisis all its own, however, often presented as a shortage of energy. This portrayal is based primarily on the theory that much of our energy comes from fossil fuels and that we are quickly running out of those fuels.

Convincing as this theory may be, it is somewhat of a mischaracterization of the energy industry's real crisis. Presenting the opportunities for nanotechnology in the energy sector as largely defined by supposedly dwindling petroleum reserves is likely to lead to an underestimation of the opportunities and perhaps to misunderstandings about what those opportunities actually are. Instead, the position that I am going to take throughout this chapter is not so much that nanotechnology can help us provide new and better sources of energy, although, in fact, this is the case, but rather that nanotechnology can provide us with new ways to concentrate energy and deliver it to the places where it is needed.

There are a couple of reasons for taking this approach. A little careful thought suggests that there is actually no energy crisis as such. There are huge amounts of energy to be had from air, wind, sea, and sky (meaning solar energy). In fact, this is pretty much the view taken by all professional commentators on energy issues.

Where these experts vary, of course, is in which source they would prefer to use, based on trade-offs in cost of extraction, environmental damage, costs of deployment, and such. Typically, these experts see the trade-offs very differently. In discussing the United States' energy issues, one journalist may point to the fact that the United States has about a quarter of the world's coal reserves and suggest that more should be done to replace oil extracted in politically unstable regions with coal. That journalists may go further and point out that new forms of fuel are appearing that can be extracted from coal using new processes, some of which, by the way, involve nanotechnology. Another expert may tell us that our continued dependence on oil is not a problem and that new extractive processes and fuel additives, again, often involving nanotechnology, will make oil reserves extend into the indefinite future.

And yet another commentator, a "green" or environmental one, will say that none of the above makes much sense because of the huge environmental impact of using fossil fuels of any kind. A much better way to go, he or she will inform us, is to use alternative energy sources, such as solar power, wind power, geothermal power, and so on. In debates, our green commentator will almost certainly be challenged by more fossil-fuel-friendly experts to come up with a way that his favored technologies can actually be improved so that they can make a serious dent in the use of fossil fuels, since many of these alternative energy technologies have been around for decades and, while profitable for some, have proved "nichey," at best from an overall perspective. In many cases, the response to this objection will be that while this was a fair point in the past, alternative energy sources are getting better. Yet again, as part of the proof that this is so, nanotechnology is sometimes cited as a major enabling technology.

This analysis suggests that nanotechnology is in a very good strategic position in the energy sector, seen as a friend to all. Therefore, it appears that the nanotech industry can serve as an "arms dealer," supporting all sides in the energy wars. This is a particularly attractive place in which to find oneself, since it is quite difficult to decide who is going to win in these wars, and nanotech benefits regardless of who wins. On the surface, everyone's arguments seem impressive. We are going to bask in an ecotopia brought about nanotechnology, which sounds good to me. We are going to go on doing business as normal with oil being the major source of power for our personal transport, because nanotechnology will make our vehicles that more efficient and that sounds good to me, too All sides of the argument are able to marshal enough "facts" to make it sound as though each of one's favored scenario is a sure bet.

In practice, however, this isn't good enough. Few firms are capable of planning nanoenergy products for *all* possible futures, and I believe such firms must perform two kinds of analysis. The first is political analysis; the second is a particular kind of economic analysis, based on the how the history of the energy industry has tended to evolve.

It behooves any nanotechnology firm that is betting on a particular energy policy to make some serious political calculations about whether the general direction of politics in both its home country and internationally favors the adoption of that energy policy. General political trends can ultimately impact energy policy and therefore need to be watched. If a huge Republican majority was swept into office in the United States on the basis of "family values" issues, this might end up favoring the fossil fuel sector. If Greens win seats in European parliaments on the basis of an antiwar stance, this may end up giving weight to solar and wind power in their energy policies. There's a great deal more to political analysis, of course, but most of it lies well beyond the scope of this book.

Economic Analysis of the Opportunities for Nanoenergy Firms

The politics of energy isn't everything of course. It seems reasonable to assume that where an energy technology has better economics than another technology it will ultimately win out. However, again it is far from clear what it means to have better economics. For the purposes of this book, however, I am going to adopt the approach argued for in a book by Peter Huber and Mark Mills, *The Bottomless Well.*[72] This book goes into considerable depth in an analysis of energy economics that stresses that it is not so much the cost of actual energy (or fuel) that matters, but rather the costs of turning it in to efficient *power* (rather than energy), which means turning the available energy into a form that can actually do useful work and delivering it to the right place at the right time.

This means that a lot more is involved than just energy generation, energy must be changed into different forms, stored until needed and then transported efficiently. For example, there is a lot of solar energy around, but what we actually pay for is the cost of turning it into useful power, which can be considerable, especially if the cost of real estate for deploying the solar panels. In fact, Huber and Mills make the case for the real estate cost often being almost ridiculously large, noting, " . . . to power New York with PV, you would have to cover every square inch of the city's horizontal surface with wafers—and then extend the PV sprawl over at least twice that area again." Whether you agree with the particular numbers that these authors comes up with or not, the real issue is how expensive it is to supply power using PV, not the fact that solar energy is in some, and undisputed way, free.

Thus one opportunity for nanotechnology would clearly be the efficiency of solar cells so that we don't have to pave New York (or Los Angeles, or California) to get the power we need. Even where there is room to create a PV power station, such as in the middle of an Arabian desert, for example, nanotechnology might contribute by reducing the cost of the individual cells. New business revenues would also flow to nanotechnology firms that can come up with better ways of storing the power from PV installations, since there are obvious and predictable variations in the output of this power throughout the day and throughout the year.[73]

More generally, in this book, we will be primarily concerned with how nanotechnology is changing the economics of power, rather than simply changing the cost of fuel (although there is some of that too.) This translates into a broader range of opportunities for nanotechnology than might have been perceived if a purely fuel based-analysis had been applied. In Table 4.1 I set out where the generic opportunities for nanotechnology will be found in the energy sector.

Table 4.1
Opportunities for Nanotechnology in the Energy Sector

Position of Opportunity in the Value Chain	Likely Contributions of Nanotechnology
Extraction	Nano-enabled enhancements can lead to reductions in the cost of extracting fuels from fossil deposits thereby increasing the available reserves—nanocatalysts and nano-enhanced drills are two areas where there is some obvious potential here. Nanoengineering may also provide better ways of harnessing renewable energy sources, especially better materials for windmills, solar power collectors, etc.
Transformation	Transformation is all about generating useful power from raw energy. This may mean simply making more efficient/ less polluting hydrocarbon fuels from raw fuels—using nanocatalysts, for example. But increasingly the opportunities will be in improving the transformation of "natural energy" sources into electrical power (i.e., electricity.) Nano-enabled PV cells are paradigmatic here. Electricity appears to be the most useful form of power known and it is important to remember that for all the talk of "solar" power or "nuclear" power, much of the debate is actually about how to generate electricity.

Table 4.1 (Continued)

Storage	The whole energy sector is beset by the problems of peaks and troughs, both on the supply side and the demand side, therefore buffers are an important factor in cost effective power systems. Electricity is stored in batteries and nanotechnology is certainly making contributions to enabling more efficient battery technology. Better batteries would also be an important factor in the future of alternative energy forms such as PV and wind power that generate power only at certain time of the day or year. But also included here would be materials for passive solar buildings that store energy from the sun in thermal form. Yet, another aspect of storage is nano-enabled gas storage, which may be particularly helpful in the much dreamed about hydrogen economy.
Distribution	One of the reasons why electricity is so widely employed is that it is so easy to transport. National and regional power grids are proof positive of this. Compare transporting huge quantities of natural gas on superhighways! However, long-haul transport of electricity is not outstandingly efficient and there is much room for improvement. Highly conductive nanomaterials, especially carbon nanotubes, offer hope for the future here.
Usage by consumer	Nanomaterials can help in a number of ways to reduce the costs of energy at the consumer level. At least initially these are likely to be fairly unremarkable in nature. Better insulation using nanomaterials would be one example. Additives to fuel oil that make them produce more energy per unit volume or mass is another. High-efficiency heating systems using nanomaterials is yet another. Nanosensors could also play an important role in conserving energy in buildings, by providing finely tuned power monitoring and control.

The Impact of Nanotechnology on the Energy Sector

As we have seen there seems to be a general agreement that nanotechnology will have a big impact on the future of the energy sector, even if no one can quite agree on what that impact (or what that future) will be. Nonetheless, based on the technology directions currently being taken by both R&D efforts and corporate commercialization programs, it seems reasonable to assume that the impact can be categorized into five reasonably well defined headings:

1. The nano-enhanced fossil fuel sector;
2. Fuel cells and the nanoengineered hydrogen economy;
3. Nanosolar power;

4. The nano-enhanced electricity grid of the future;
5. Nanopower for the pervasive communications network.

The vast majority of nanoenergy businesses will find that they fit pretty well into the areas listed above, which also overlap each other in some ways. The end game for the nano-enabled energy sector could be a dramatic change in the world energy picture, with major disruptions in the kinds of energy used by industry and consumers and opportunities emerging for those currently without access to reliable energy sources. In a book of this kind, it is not possible to explain all the nuances in what is an extraordinarily complex situation involving geopolitics as much as it does technology. However, one important fact to note is that nanotech potentially alters the geography of energy, making the Europe, North America, and Japan less reliant on foreign oil, and making certain alternative energy technologies more widely available than might otherwise have been the case. In addition, one has to understand the interplay of developments from one area to another. For example, the economics of natural energy sourcesis affected by the storage technologies available. This is but one example of a broader ability of nanotechnology to shift boundaries in the energy industries, once thought to be unmovable. Above all, nanotechnology seems set to provide new approaches in both the fossil fuel and alternative energy sectors that will transform the economics of both and will greatly affect the balance between them. This likely to confound those who say that there are currently no alternatives to fossil fuels, and this will continue for many decades to come. It is just as likely to confound those who want us to believe that only a wholesale move to alternative (and supposedly sustainable) sources of energy can save us from an ecotopia.

The Nano-Enhancement of Fossil Fuels

Even the "greenest" commentator on the energy scene will admit that we will rely on fossil fuels for the foreseeable future. As we will see, most of the alternative energy sources do not have the efficiency to produce power at a cost, or anything close to the cost of fossil fuels. At best, these alternatives are competitive with fossil fuels only in limited geographies. For example, most of Iceland gets its energy from geothermal sources, which, because of the volcanic nature of the Icelandic geologic, is plentiful. This is an interesting, but highly unusual situation. Even Jeremy Rifkin, one of the public intellectuals most vociferous in his support of alternative energy sources, notes in his book *The Hydrogen Economy*, that, "PV power is still two to five times more expensive than conventional electricity generated by fossil fuels," although he is much more optimistic than Huber and Mills on the long-term potential for PV.

Given all this, it seems reasonable to assume that that the high degree of reliance that the world currently places on fossil fuels is likely to persist for many years to come, unless, of course, nanotechnology proves to be more effective than even its most ardent supporters expect in tipping the balance towards alternative sources of power. Meanwhile, the pressure on fossil fuels is being increased by the growth in demand for energy, especially from the so-called lesser developed nations, which, in many cases, are not as undeveloped as they were when that term was coined. Current global energy consumption is around 10 billion tons of oil equivalent a year and rising at around 2 billion tons per decade. By 2030 developing countries are expected to be consuming close to half of this energy, compared with the current 30%. The high growth rates of China and India, each with an enormous population, are often seen as primary contributors to impending energy problems.

Where nanotechnology is most likely to have an immediate impact is on reducing the cost of power provided by fossil fuels. This is where I believe the most immediate opportunities for nanotechnology lie. As I have already emphasized (as shown in Table 4.1), this is not just a matter of reducing the cost of extraction oil, natural gas, and coal. Instead, the opportunities are broader, stretching down the value chain from the oil well or coal mine to the gas pump or coal/gas delivery truck and on to better insulation and improved boilers and furnaces in homes, offices, and factories.

Unfortunately, there is not enough space in this book to go into all the areas where nanotech will affect the economics of fossil fuels. Instead, I want to highlight a few areas that seem to me to be particularly important, either because they are likely to be significant revenue generators in their own right or because they are pointers to how nanotech may become commercially important in the future, or both. Most of these applications may seem rather mundane compared with, for instance, nanomedical wonders that I discuss elsewhere in this book. However, it is important to remember that without cheap and plentiful power sources, most of the other, "sexier" applications for nanotech simply wouldn't exist. It is also important to remember that many of nanotech applications profiled in this section represent real short-to-medium term revenue streams for firms with the resources required to exploit them. Nanocatalysts are already being used to improve power performance of fossil fuels. By contrast, nano-enabled medical miracles are likely to be delayed by the regulatory requirements that are part of the business environment in which the pharmaceutical and medical device industries operate. Therefore, in the long run, the opportunities for nanotechnology in the energy sector are potentially very dramatic indeed.

Nanocatalysts One aspect of nanotechnology that is especially important in the context of the energy sector is catalysts. A catalyst is a material that speeds up a chemical process without being itself changed in that process and catalysts have long been a stock in trade for the chemical industry. What catalysts do, in effect, is to reduce the amount of energy needed to carry out a given chemical process for a given amount of material. Probably, the most familiar example of a catalyst in action for many of us is the catalytic converter in a car. In this system, platinum acts as a catalyst to improve the conversion of the dangerous gases carbon monoxide and nitric oxide to carbon dioxide and nitrogen. The physics of the platinum is such that the nitrogen and oxygen atoms bond separately to the platinum atoms, the nitrogen atoms then combine to form nitrogen gas, which floats off into the air (air is mostly nitrogen anyway). The remaining oxygen atoms combine with the carbon monoxide to form carbon dioxide, again a major constituent of the air we breathe. The platinum remains unchanged.

The same effect could be achieved through some kind of thermal process, but then a lot more energy would have been consumed. With catalysis being such an important part of the energy and chemical industries, it is a reasonable question to ask as to whether nanotechnology can improve the catalysts themselves. As it turns out, there is an inherent characteristic of nanoparticles that make them more suitable than other materials to serve as catalysts, and this is their size The smaller the particle size, the bigger the surface area of the particle relative to the size. The bigger the exposed surface area the more "catalytic power," because this power is dependent on the available surface to which the atoms from the chemical that being changed in the process can be attracted. Nanoparticles have bigger relative surface areas because the surface area (S) is proportional to square of the radius (r) of the particle, while the volume (V) is proportional to the cube of r. Hence S/V is proportional to 1/r and as r gets smaller, S/V gets bigger.

The undeniable mathematical advantage of nanoparticles in this context coupled with the importance of catalysts have already lead several firms to pursue this opportunity and, in fact, catalysts are one of those areas where many catalysts already in use are actually nanocatalysts, even if they are not referred to as such. This implies that nanocatalysts fall into the category of *accidental* or *evolutionary* nanotechnology as defined in an earlier chapter.[74] Designing and using nanocatalysts, it should be noted, goes well beyond just focusing on nanomaterials. Reactants need to make their way to the catalytic site at a rate sufficient to exploit the available reaction rates, which can imply structures with mixed scales; or a complex nanosystem may be built, for example, catalytic nanoparticles with precisely controlled dimensions supported on a nanoscale structure. Carbon nanotubes sheets, for another example, would provide good electron mobility and might be very suitable in such an application.

According to one financial analyst, one impact of nanocatalysts might be that we will ultimately "30 cents worth of nanoscale nickel to replace $7 worth of platinum."[75] The importance of nanocatalysts is also enhanced by the fact that their addressable market is huge as almost all products derived from fossil fuels (as well as plastics) use catalysts in their manufacture As an example of where nanocatalysts may take us in the fossil fuel industry, NanoKinetix has announced that it has a nanocatalyst that can produce premium gasoline at the cost of regular. Several other firms have been hard at work on nanocatalytic improvements for the catalytic converters described above.

Coal Liquefaction If the applications of nanocatalysts that we have discussed to date are relatively mundane applications of nanotechnology, it should be remembered that the applications set out above are opportunities that will generate revenues more or less immediately, if they are not already doing so. A somewhat more revolutionary, if slightly more distant, application for nanotechnology in the energy sector is taking existing fossil fuels and processing them using nanotechnology into other fossil-based fuels that overall better overall cost-to-power performance. Given that petroleum oil is a fuel that is unattractive for many all-too-obvious reasons, the natural fuels for nano-engineered energy solutions would certainly include natural gas and coal. We have more natural gas left than oil and hundreds of years' worth of coal.

The biggest nano-opportunity here is liquefying coal using nanocatalysts. The end result is a clean diesel-like fuel for which it is not especially hard to adapt vehicles. The fuel is actually clean, only in the sense that vehicles burn it cleanly; the impurities are released during the production process. However, centralized production of pollutants and greenhouse gases allows for more effective containment than when they are being produced by every car and motorbike on the road, which is one of the reasons why cars powered by hydrogen made from fossil fuels is not as absurd as it might appear to some.

From a U.S. perspective, the biggest incentive for this opportunity is the fact that we have a considerable amount of coal and since the 1970s, we have been hugely reliant on the unreliable sources of oil for transport. Putting those two facts together, it would favor a technology that can turn our abundant coal into a replacement for oil, a commodity that may be difficult or expensive to get.

Coal liquefaction is also of growing interest in some Third World nations, who also have some coal, but little oil. These countries face the prospect of a huge rise in car usage in the early part of the 21st century, with a proportional rise in the need to import fossil fuel. For example, in China and India only around one in one hundred people own cars, compared with two out of three in the United States. Even a small increase in the numbers of people obtaining cars would represent a major jump in demand for transport fuel. China has taken a nanotech-enabled step in this direction. The $2 billion Shenhuan coal liquefaction project, using nanocatalytic technology developed in the United States, is expected to be an economically competitive way of producing fuel if oil dollar prices are above the low twenties per barrel. At the time China started this project, this was considered a fairly high price. At the time I am writing these words, it is hard to remember that oil was once this inexpensive.

Gas Liquefaction Nanocatalysts can also be used to liquefy gases and this technology is at a similar stage in development to coal liquefaction, which is to say it is likely to be a short-term revenue generator. However, the market drivers for gas liquefaction are a little different to those for coal liquefaction and the long-term revenue potential could be much larger, if gas liquefaction turns out, as some expect it will, to be a gateway to the much-talked-about hydrogen economy.

However, the first commercial impact of nano-enabled gas liquefaction will be not on hydrogen, but on natural gas.

If you consider natural gas purely from the perspective as a fuel, it appears to have an advantage over oil in terms of availability. At the current level of technological development, there appear to be decades more gas reserves than there are oil reserves. If deep ocean reserves of gas prove accessible, we are possibly talking centuries more of gas reserves than those of oil.

This is where the perspective of looking at power rather than energy is particularly fruitful, because the cost of gas at the source is only part of the story. Gas is inherently difficult and expensive to move over long distances, primarily because gas is—to state the obvious—gaseous. Energy economists point to significant reserves of "stranded," gas reserves that are too far away from where it is needed to be of economic value. If and when there is a general inflation in the price of power, some of these reserves will suddenly present themselves for commercialization. However, nanotechnology can also make a significant contribution by making it easier and less expensive to liquefy natural gas, thereby making it easier and cheaper to transport. And natural gas loses little or none of its power in the process of being liquefied.

As with coal, liquefying natural gas does not necessarily require nanotechnology and liquid natural gas plants are now appearing all over the world as the costs of liquefaction have been falling. Where nanotechnology seems most likely to make a contribution, however, is once again through nanocatalysts that make it easier to convert gas to an easily transported liquid. As cynics about the potential for nanotechnology will no doubt be quick to point out, the catalytic conversion to liquid of fossil fuels (both coal and gas) has been achievable since the 1920s using the Fischer-Tropsch process. The end result is something like a diesel fuel, although it is extremely clean compared with regular diesel. This is because impurities are removed at the liquefaction plant, which is easier to do than in the machinery that actually burns the fuel.

Nanotech, the Environment, and the Road to the Hydrogen Economy

Pollution is a topic that is intimately tied up with the energy sector, because most of the fossil fuels are highly polluting. Nanotechnology firms may well help in this regard, again using nanocatalysis, and again as a near-term opportunity. Nanocrystalline catalysts made from cadmium, selenium, and indium have shown to be effective carbon dioxide filters, while titanium oxide nanocrystals under UV light will remove mercury vapor. Although such approaches certainly represent nanotechnology in action, their impact pales in comparison to how nanotechnology could potentially make a dream of certain environmentalists come true, namely, the "Hydrogen Economy," which is also the title of Jeremy Rifkin's book on the topic.

The basic idea behind Rifkin's thinking is that the geopolitics, environmental dangers, and scarcity of fossil fuels are becoming just too hard for us to cope with, and we need something better. That "something" is hydrogen or, more specifically, hydrogen fuel cells. Hydrogen fuel cells use energy liberated when oxygen (from the air) and hydrogen combine to produce electricity. These fuel cells come in different sizes that could be used in cars, homes, offices, or mobile computing and communications devices. In theory the hydrogen economy would have many advantages over the fossil fuel economy. There is no such thing as the geopolitics of hydrogen because it is the single most abundant element in the universe. There are no environmental dangers associated with hydrogen (according to Rifkin) because the end result of the hydrogen fuel cell process is water and heat.

Unfortunately, while this all sounds very nice, it's not quite that simple. To employ the "power" versus energy concept yet again, hydrogen-based energy, releasing energy from the reaction between hydrogen and oxygen, is certainly a very inexpensive process. However,what really matters is what the cost will be to deliver the power from such fuel cells to the places where it is needed. First, it is not easy to store and transport hydrogen because it is a gas and that's where the gas liquefaction process described above becomes a consideration. Also, while it is perfectly true that hydrogen power would be less polluting in the obvious sense of the word "polluting," the ubiquity of a highly flammable gas that could (and almost certainly would) leak out would pose a safety hazard, although again, liquefaction may at least reduce the safety risk to the current one associated with gasoline. It is an interesting and open question as to whether the water vapor released into the atmosphere by hydrogen cells could also prove a climate change agent. Other problems exist, in that hydrogen will have to be harvested from other gases and fuel cells will have to employ new technology to make them much more efficient if they are ever to be more than just a niche product.

Nanotechnology looks like it may be able to tackle all of those worries in one form or another, and a lot more is involved than just nanocatalysts and liquefaction. As we show in Table 4.2, nanoengineering seems to offer solutions to almost all of the problems that seem to be holding back the hydrogen economy, which incidentally has received considerable government support as well. This is not to say that the hydrogen economy is a sure thing thanks to nanotechnology.

Table 4.2
Nanotech's Impact on a Future Hydrogen Economy

Issue	**Problem**	**Nanotech's Contribution**
Harvesting hydrogen	Hydrogen needs to be extracted from other gases such as natural gas and methanol, as well as from gasoline. Current means of creating hydrogen can be polluting.	This can be accomplished better with nanostructured membranes and nanocatalysts, which reduce the cost and increase the efficiency of the process. In some processes nanocatalytic effects are used to break up water into its constituent hydrogen and oxygen atoms.
Storing hydrogen	As a gas, hydrogen takes up a lot of space relative to its energy density. Storing a highly flammable gas also raises obvious problems.	Liquefaction as described in the main text is one possible way to cope with storage problems. In additions using nanostructured materials than are highly absorbent of hydrogen would increase the efficiency of hydrogen storage.
Transporting hydrogen	Flammable gases are expensive and dangerous to carry.	Liquefaction is a partial answer to these problems.
Electricity generation	Fuel cells are not very efficient.	Nanocatalysts can release energy with greater efficiency.
Other safety issues	Leaks and fires.	Nanosensors could be deployed to alleviate this worry. Hydrogen nanosensors are expected to be used in many fuel cells.

Seeing how nanotechnology can provide broad based solutions to creating a hydrogen economy is one thing, but actually implementing those solutions is quite another. Fuel cells cars have been promised for a long time, but have never proved very commercially viable. Nonetheless, the possibility that nanotechnology could fundamentally transform the economics, politics, and technology of power is a real one and a wonderful example of the revolutionary nature of nanotech. Or, put in more mercenary terms, nanotechnology plus hydrogen may well create huge new opportunities for business people and investors in this new century.

Even if the developments outlined in Table 4.2 don't end up being the definitive path to the hydrogen economy, they may well provide important new business directions for chemical and energy firms. For example, the nanomembranes that are being proposed for harvesting hydrogen would have broader uses. Some researchers working in this area believe we are not far off from being able to use nanoporous membranes to separate nitrogen from oxygen, two molecules of very similar size currently separated in industrial quantities using energy-intensive cryogenic methods. Such separation technologies also hold out the promise of improved CO2 separation; which would be of importance to those interested in minimizing the impact of greenhouse gases, assuming that this continues to remain a policy fashion.

I'll finish this section with a discussion of nanotechnology's potential contribution to storing hydrogen. This is of importance in the context of fuel cells, but also a key factor in any hydrogen driven economy that may emerge. Just imagine the fossil fuel economy without an effective way to store oil, coal, and natural gas. Of course, hydrogen could be stored in vast tanks, but the emphasis here is on the word "vast," because the energy density of hydrogen is quite low. To effectively power a car with liquid or gaseous hydrogen using conventional technology you would need a tank that was basically too large to carry on the car you were trying to power. By extension, if you were trying to power an entire advanced economy, much of the real estate in that economy would have to be turned over to fuel tanks. An additional problem for hydrogen is leakage, obviously a problem with any gas storage, but especially bothersome with hydrogen, because its molecules are so small and they can wiggle through the crystalline structures of the metals used to build the tanks in which they are stored. This leakage is exacerbated by the fact that hydrogen is stored under pressure in both industrial and automotive fuel cells.

Nanomaterials may come to the rescue here in a number of ways. Nanomaterials could find a use in creating less permeable storage tanks capable of holding hydrogen at higher pressures. Another likely innovation is a "hydrogen sponge" that will soak up hydrogen in much the same way that a real sponge soaks up water. While the purpose and science behind a hydrogen sponge is quite different to that of a nanocatalyst, the "nanogeometry," is quite similar. Both nanocatalysts and nanosponges exploit the fact that in nanostructures, surface areas can be huge compared with the volume of materials deployed. Translated into the context of a sponge, this means that in a moderate sized sponge there are lots of holes for the hydrogen to sink into and hence storing enough hydrogen for practical uses becomes possible.

Materials capable of making hydrogen sponges have been around for a long time in the form of metal hydrides. However, they do not store enough hydrogen to be useful unless they are heated to 250°C or so, which make them impractical for purposes outside of R&D labs. This is yet another of the many areas mentioned in this book, where carbon nanotubes (and potentially other nanofibers) seem destined to play a role, since they seem to be able to absorb enough hydrogen at room temperature to be useful. (The U.S. Department of Energy says that a nanosponge will need to store at least 6.5 percent of their own weight to make fuel cell based cars practical. Carbon nanotubes seem capable of 8 percent storage by weight, if metallic nanocatalysts are also utilized.)

Hydrogen storage seems to be the Holy Grail for researchers and firms backing the long-term shift to hydrogen based economy and there are lots of projects in this area being carried out in universities and in private labs. A breakthrough using nanomaterials would be one of the more spectacular successes for nanotechnology in the near term, and would be accompanied, at least in the United States , by the kind of financial rewards that one has come to associate with this kind of breakthrough. However, a few words of caution before I close on this subject.

First, announced breakthroughs in the energy sector are sometimes not all they are cracked up to be. Many of the readers of this book will remember all the fuss about cold fusion, which ended up being one of a long line of energy technologies that was going to make electricity too cheap to meter, but failed to meet its initial promise. Something similar to this also occurred back in the late 1990s, when a team at Northeastern University claimed to have created nanomaterials that could store up to 65 percent of their own weight. This could have propelled the world quickly into a hydrogen age, but the research team's findings could never be reproduced. In addition, what we have talked about here so far is a material that is more efficient than previous materials for storing hydrogen. However, what will be actually required in the marketplace is not a nanostorage material, but a nanostorage *system* for hydrogen. This implies ensuring that hydrogen can be stored and downloaded quickly and efficiently and that the lifetime—or time between charging—of the storage material is suitable for the application.

Nano-Solar Power

Like fuel cells, solar power has been on the verge of solving our energy problems for a long time, but somehow has never managed to actually do so. In part this is because both hydrogen and solar power share the fact that they are based on inexpensive, but low energy density fuels, so by the time they are delivered to the customer they are quite expensive. The promise of nanotechnology in both cases is that it will make both solar and hydrogen power much cheaper to do useful work.

There are actually (at least) three ways in which people have proposed harnessing the power of the sun:

- *Passive solar,* in which building materials, architectures, and geographical architectures are designed in such a way that energy from the sun is utilized to keep the house warm. Passive solar heating involves two main elements: south facing glass and a thermal mass to absorb, store, and distribute heat. It seems possible that nanotechnology could make a contribution here with better materials to improve both aspects of a passive solar power heating system. Passive solar has proved quite effective, although it is not a system that can be easily retrofitted, because much of its effectiveness lies in the basic design of the home or other building.
- *Solar power stations* are like all other power stations turbines that create electricity through an electromagnetic effect. In most power stations the turbines are driven by fossil fuels (oil or coal mostly) and some power stations are driven by heat from nuclear power. In solar power stations, the turbines are driven by heat from concentrated solar power. This is a fairly unusual way to generate electricity and perhaps nanotechnology contribution to making solar power stations more widespread with some kind of nanostructured solar collectors, but it doesn't seem to be much of a priority for nanotechnologists at the present time.
- *Photovoltaic (PV) systems* are systems that come in various shapes and sizes that create electricity through the photovoltaic effect, which is a physical phenomenon in which electrons are freed from materials by bombarding them with photons. In this case the photons are coming from the sun. Photovoltaic systems have been around for quite some time and have found a number of niche applications, although they never seem to have lived up to the lofty expectations of some of their backers. Several companies are researching how nanoengineered

> materials can reduce the cost of photovoltaic systems. This is important because even though the input to photovoltaic systems is free energy from the sun, the systems are typically expensive, so that the cost of electricity generated in this way is significantly more costly than electricity generated by other means. In addition, the conversion of solar energy to electricity in photovoltaics is highly inefficient, meaning that—as I have already mentioned—a lot of real estate must be taken up by solar cells to provide much energy.

For the rest of this section, I am going to focus on PV systems, since this is solar power type that is attracting the most attention as far as the nanotech community is concerned. As we have already noted, the fuel for PV is free, but the capital costs of the PV system are high and the efficiencies are low, so to obtain large quantities of power a considerable amount of real estate is taken up with solar panels. As a result of these undeniable facts about solar, the prospects for solar energy have often been dismissed by analysts, who also note that the improvements in cost and efficiency of PV have not changed much in 30 years. Nanotechnology, however, may be able to waken PV from its slumber.

The economics of traditional PV has been built around a materials platform that is based on the same silicon technology used for computer chips and, as I have already noted, these economics are not especially compelling because of the high up-front costs associated with PV. The magnitude of these costs is compounded by the need for storage, again, much like hydrogen. You can't switch the sun on just because you want to bake a cake or bathe the kids. So solar energy from PV needs to be stored so that it can be used when needed. Batteries cost money and take up space. This means that the future of PV is intimately tied up with higher energy densities, which is something that nanotechnology can bring about (see below). Nanoengineered improvements in fuel cells can help too, since solar energy might be used to make hydrogen as a storage medium.

The high up-front costs of PV are a serious marketing impediment in and of itself because many homeowners unwilling to make the upfront investment necessary to save money with PV, even if those savings were well established. There are many industrial users and some homeowners who crunch the numbers and find that the extra initial cost is worth it because of the payback in terms of fossil fuels not used and perhaps in terms of longevity of the equipment. But there are not enough of these to make for truly widespread adoption of solar in the sense that the various fossil fuel based systems are widely adopted.

Where nanotechnology can help, in addition to the storage issues raised above, is either by increasing the efficiencies of PV cells or reducing the cost per unit of area of the PV cells themselves.

The current efficiency of PV is about 15 percent, but efficiencies of around 30 percent have been seen in the lab for a long time and by exploiting more wavelengths available from the sun, it is possible to push this efficiency up past 60 percent. If deployed commercially this means that roughly one third of the PV cells currently used would now need to be used for a particular application. However, these new highly efficient PV cells are likely to be more expensive than regular cells. Much of the work in creating highly efficient PV cells does not involve nanotechnology, but rather the use of new compound semiconductors—typically, gallium compounds instead of silicon. However, lead selenide nanocrystals serving as quantum dots have been shown to increase the efficiency of PV cells to 65 percent. This has been demonstrated at the U.S. Department of Energy's National Renewable Energy Laboratory (NREL).[76]

Where nanotechnology seems closer to commercial reality in providing better economics for PV is in the form of low cost solar panels, that reduce the initial costs associated with PV, but at the expense of efficiency. The materials platform for at least some of this technology is based on titanium oxide nanocrystals, but my experience at NanoMarkets is that many of the firms active in this space are being *especially* secretive about their "secret source."

The first products that are likely to appear using this technology will have efficiencies well under 10 percent, but it has been claimed that they could reduce the cost of PV by as much as 90 percent. However, this is probably an exaggeration, because the new breed of low-cost solar panels seems likely to have a shorter lifetime than conventional panels. In fact, the low efficiency of these new PV cells is likely to make them less of substitute for conventional PV than a new kind of product altogether. Indeed, with many venture backed firms, such as Konarka, Nanosolar, and Nanosys, and some other very large electronics firms well advanced in this technology, it may be the biggest business opportunity in nanoengineered PV right now is finding profitable new applications for them that are "out of the box" from the point of view of the traditional PV industry.

It seems to me that PV is likely to be used initially in the form of auxiliary power sources than as a primary power sources in various circumstances. There has also been talk of inexpensive roll-to-roll printing of solar panels, using low-cost PV to keep the batteries of notebook computers and cell phones charged and even painting PV cells on the walls of office blocks. (The last of these suggestions somewhat avoids the problem of low-efficiency cells being impractical because of the all of the real estate they take up.) Efficiencies of this type of product will certainly increase and they appear to have the potential to take PV in directions that it has never before been emboldened to go. Also, the efficiency of nanoengineered PV may be lower than standard PV, but in some formulations, energy is retained longer so that "nano-PV" actually has better performance indoors and in other low light conditions.

The Nano-Enhanced Distributed Electricity Grid of the Future

The economics of any networked distribution system is a constant trade-off between using numerous hubs linked by short distances and a network in which a few large and powerful hubs linked by large distances. This trade-off is made by balancing the cost of transmission against the cost of the hubs. Much of the history of the telecommunications industry in the past 40 years could be written in terms of the shifting architectures as new technologies moved the balance from a few big switches to many small switches and back again.

Electricity generation and distribution, like telecommunications, is a networked industry, but it is one that, until recently anyway, has not seen huge amounts of technological change and the balance between "hubs" and "transport" has almost invariably favored large hubs, that is, large turbines. There are significant economies of scale to be had from big power stations. Primarily, the cost per kilowatt of power generated by big turbines is typically less than that generation by smaller turbine. In addition, it is usually more cost effective to deal with waste and pollution from a single centralized location than from many different locations.

As was the case with telecommunications, the bias towards large hubs could only change if the relative cost of small hubs came down or if the relative cost of transmission went up, which is essentially the same thing. Note that if one could bring down the cost of transmission using nanotechnology or some other method, that might be a good thing in terms of reducing the overall cost of electricity infrastructure, but it would clearly work towards retaining the status quo of large generators.

As it happens, nanotechnology could push the electricity industry in both directions over a period of time. First it may lead to improved economics for smaller generators, leading to a more distributed grid. Then at a later date, it may lead to reduced transmission costs, which work in favor of large turbines. One immediate opportunity seems to be to use nanotechnology to lower the cost of smaller nodes, making for a more distributed network. A lot of development has been seen in "miniturbines," using the same principles as larger power stations but on a smaller scale. These machines compete directly with larger fuel cells for small-scale industrial use and nano-enabled fuel cells should also be seen as a way that nanotechnology encourages the distribution of energy generation from where we are now. For miniturbines, the impact of nanotechnology is not likely to be revolutionary but there are certainly applications for nanocrystalline metallics, ceramics, and composites that can improve performance parameters, especially lifetime.

Working in the other direction, towards a more centralized power system, is the possibility that highly conductive nanowires, nanotube composites and nano-enabled superconductors would vastly improve the efficiency of the electricity grid. You could then have a relatively few power stations generating power and then distributing them cheaply across the country or the world. Large nanoenabled batteries, which would likely again use nanocatalysts, would vie with "supercapacitors" built from nanomaterials to provide electricity storage at centralized power stations, which would become more efficient, because they could make electricity at times when the demand for electricity was not so high and use it at times of peak demand. However, it is fair to say that nano-enabled storage solutions would also play an important role in the more distributed scenario for electricity distribution. To illustrate the point, imagine how much a PV-powered home or office would benefit from an inexpensive and efficient means of storing electricity for hours and days when the sun was not shining.

It was suggested by the late Professor Richard Smalley, the discoverer of the buckyball, that the implementation of new nano-enabled electricity grid should become a matter of national priority and that something like the Kennedy space program should be launched to support it. The economies of large-scale production coupled with low-cost distribution would make us much less dependent on fossil fuels, it is believed by those who advocate this kind of solution. However, at the present time, it seems most unlikely that the United States, or any other major government, is going to invest money in this kind of program. Whatever the energy policies of the United States and other governments, the technology for this kind of project is still many years, even perhaps a decade or so, from commercialization. One factor that needs to be taken into consideration in particular is that reengineering the grid along the lines suggested by Smalley would take not just a lot of capital, but also a lot of nanomaterials, which in turn would require a manufacturing infrastructure that does not now exist and would take years to develop.

Nano Power for the Pervasive Network

I conclude this chapter with some brief notes about a different kind of "energy crisis" than the one generally associated with this term. I refer to the energy crisis that faces the mobile communications and computing sector. Basically, the cause is that Moore's Law is letting the suppliers of cell phones, notebook computers, PDAs, and such add more and more features, but the energy density of the lithium ion batteries used to power these mobile devices have been increasing at a much slower rate. The end result is the time between charges of the batteries is actually beginning to decline. According to a recent buyer's guide publication from *Consumer Reports*, the average time between charges for a notebook computer has declined to about 2.5 hours.[77] Nokia supposedly the abandoned the launch of a new many-functioned cell phone, because its functionality put so much demand on the battery that it needed constant recharging.

This is a big deal. As I have mentioned elsewhere in this book, many leading electronics and computer firms are pushing the idea that the next big thing in network will be a pervasive network of wireless devices capable of video, data, and voice communications. If this vision of the next generation of networking is to pan out, there will clearly need to be better ways to power the next generation of mobile devices and the generations that follow. This means either improved battery technology or the use of fuel cells or solar power, either instead of batteries or as a supplement to battery power.

With regard to solar power, we have already explored the ways that nanoengineering is enabling low-cost (and low-efficiency) PV. It is easy to see, by following the ubiquitous example of calculators, how PV could be used to power cell phones, notebooks, and the like. Indeed, a number of magazine articles seem to have presented the potential for this new generation of PV in this way. Having investigated the situation fairly closely and talked with some of the firms involved, my firm NanoMarkets has come to the conclusion that PV-powered cell phones are little further off than a casual glance of the literature may suggest. First, the technology as a whole is probably still a couple of years away from commercialization. Second, the low efficiency of the current nano-enabled PV means that at best it will initially be used for providing some kind of a boost to the main battery, rather than being the main source of power, for at least a few years to come.

A very similar story can be told about fuel cells. Again these are sometimes presented as the solution to the mobile power problem. In particular, the direct-methanol fuel cell (DMFC), is said to be capable of leading ultimately to a cell phone that needs recharging only every two weeks (or a laptop that can run all day on a single charge) and that can be recharged with a brief squirt of fuel from a pressurized canister. NanoMarkets' market research has indicated that if this kind of scenario ever comes about it will be a long way in the future.

Although fuel cells for laptops are already commercially available they are more of a product for "geeks" than a serious solution for mobile IT. For one thing their energy density has to be improved to a point where they are not so bulky that they would be inconvenient to carry around. Secondly, a worldwide distribution network needs to be set up so that standard fuel canisters are widely available. Right now, my laptop may run down every three hours, but I can plug it in just as easily in Birmingham, England as in Birmingham, Alabama. However, searching for a store where I can buy just the right kind of gas canister for my particular fuel cell could prove quite daunting.

It seems likely that the fuel cell's problems can be sorted out. As we have already seen, nanotechnology is likely to provide a boost in the energy density of fuel cells and it is not beyond the imagination that a distribution network could be set up for the wide variety of fuel cell refills that are likely to emerge, since similar networks exist for printer cartridges and copier toner. But NanoMarkets research indicates that the first mass market fuel cell products for the notebook (and perhaps cell phone) market are likely to come in the form of a portable fuel cell charger for batteries, rather than a replacement for batteries. And even when fuel cells reach a level of miniaturization and power density that enables them to be the main power source for mobile devices it seems likely that they will still need a battery somewhere in the system to act as a buffer, when the equipment being powered needs a particularly large amount of energy.

All of which means that we are stuck with using batteries in mobile devices for the next few years. There are, however, some ways in which nanotechnology can help to improve the performance of lithium ion batteries. Most notable of these is to replace the conventional carbon anode that is used in lithium ion batteries with nanomaterials that provide a bigger surface area for electrons to collect on, which translates into more power from the same sized battery, faster charge times, and a longer-lived battery. Among the firms that have done work along these lines are Hitachi and Altair (which uses lithium titanate). Another firm, mPhase Technologies, is extending the time between charges by using carbon nanotubes to act as a barrier, ensuring that no chemical reaction occurs when the battery is not in use, as typical lithium ion batteries have some chemical reaction going on constantly even when not in use.

Summary: Key Takeaways from This Chapter

The number of ways that nanotechnology can impact the energy industry is very large and I have certainly not covered all of them in this chapter. What I have attempted to do, however, is to outline the main changes that are likely to come about in the energy sector as the result of nanoengineering over the coming decades. A summary timetable of the areas that we have reviewed appears in

Table 4.3. What I have indicated is that nanotechnology will shift the balance between the traditional energy sector, built on the fossil fuels that drove—in literally every sense of the word—the advanced economies of the 20th century and

Table 4.3
Timetable for Nano-Enabled Energy Solutions

	Opportunities		
Materials platform	2005–2007	2007–2010	Beyond 2010
Fossil fuels	Coal and gas used, liquefied with nanocatalysts, used as an oil substitute for transport and (perhaps) power stations.		
Hydrogen	Nano-enabled fuel cells used for charging mobile phones and notebook computers and for some niche stationary applications.	Fuel cells widely used in mobile applications and for local distributed power generation.	Nano-enabled harvesting and storage a significant commercial operation. Fuel cell driven transportation.
Solar power	Nanomaterials may make a contribution to passive solar power.	New generation of low-cost nano-enabled PV cells may be used to boost power in mobile devices and in building materials of various kinds.	Higher efficiency nano-enabled PV panels, perhaps employing quantum dot technology, helping to spread PV technology, both geographically and into different product sectors.
Batteries	Nanomaterials used on electrodes that improve the power density and charging characteristics of batteries.		
Electrical transmission			Nanomaterials provide highly efficient transport of electricity and create a vastly improved grid.

the alternative energy sources that have been much talked about but have largely been sideshows to the main fossil fuel big picture.

How the balance between traditional and alternative energy sources will be changed by nanotechnology is quite hard to say at the present time, since it will make deriving power from *both* fossil and alternative fuels better. What seems likely is that, thanks to nanotechnology, alternative power sources will encroach on the fossil fuel sector. That nanotechnology will be the killer technology, as it were, that propels the world economy out of the age of coal, natural gas, and oil and into the hydrogen economy, seems unlikely to me, but it would be very easy to be wrong about such a forecast. Ray Kurzweil has written that as the world moves to nanomanufacturing, the need for energy will radically decrease, a nice idea, but one that's hard to fully accept right now. In addition, some alternative energy sources seem just plain wrong for certain applications and nanotechnology seems unlikely to change that. The number of wind-powered cars that appear on the market in the next couple of years will be zero. We know of no development in nanotechnology that is likely to change that.

Nonetheless, changes in the world of energy can be rapid and dramatic and it seems highly likely that evolutionary and revolutionary developments in nanotechnology will be at the heart of many of those changes from now on. Firms and individuals looking for nanotechnology opportunities in the energy sector should remember a few central themes in analysing potential new businesses:

1. The appropriate way to analyze the impact of nanotechnology on the energy sector is not just in terms in terms of the cost of fuel, but rather in terms of power, that is liberating the energy from the fuel and getting it to the right place at the right time.
2. One of the most important opportunities for nanotechnology in the energy sector is the nanocatalyst. This is not revolutionary nanotechnology, since catalysts are already widely used and some catalysts already in use are in effect *nano*catalysts. What is changing is that nanotechnology means that nanocatalysts can now be more easily designed to fit specific needs. Why are nanocatalysts special? Because they expose more surface area per unit of volume and it is the surface area of particles in catalysts that drive their power. Nanocatalysts can be use to create better forms of fossil fuels from coal, oil and natural gas and to liquefy gas, this latter being an important step on the way to the much fabled "hydrogen economy."
3. Various firms, both large and small, have been attempting to commercialize fuel cells for years, but with relatively little impact.

Nanotechnology may well prove to be the key enabling technology that really moves fuel cells into widespread commercialization. This is because nanocatalysts, nanoengineered membranes, and nano-enabled generation, distribution, and storage of hydrogen seem capable of transforming the economics of the "hydrogen economy."

4. Photovoltaics has found many niches in which it can profitably be used, but its widespread use—in the sense that fossil fuels are widely used—has been frustrated by high up-front costs and relatively low efficiencies. In the laboratory, efficiencies have been quadrupled using new materials (not necessarily nanomaterials) and quantum dots. Meanwhile, a new breed of very low cost (but low efficiency) nanoengineered solar cells are about to hit the market for use in various auxiliary power functions.
5. Nanotechnology seems set to change the architecture of the power grid over the next few decades. It will help create the economically viable small-scale generators and fuel cells that will lead to a more distributed network and away from huge power stations. But there is a very long-term opportunity to create highly efficient transmission of power with power lines built with nanomaterials of various kinds. This would take both a huge investment and the availability of nanomaterials in quantities not yet commercially available.
6. Nanotechnology also seems capable of producing a partial solution to one of the thorniest problems in pervasive computing, namely sources that can provide enough power to keep a mobile device going for many hours. Although both PV and fuel cells offer something in this regard, they simply don't have the energy density to replace batteries, but they may offer some interesting solutions in the form of auxiliary power and battery chargers. Meanwhile, the ubiquitous lithium ion battery is being remade with nanotech, using improved electrodes that provide more power, longer times between charges, and faster charges when they do need to be charged.

Further Reading

Although energy issues are mentioned in most general accounts of nanotechnology, there is no general text on the subject. Worth a look is the conference report from a 2003 conference at Rice University called *Energy and Nanotechnology: Strategy for the Future.* This report can be found at http://www.rice.edu/energy/publications/docs/NanoReport.pdf. Much of this report deals with public policy issues and the idea of a national nanoenergy program being implemented in the Unites States, but it contains much that is interesting from the perspective of the business person looking for opportunities in this part of the nanotech sector.

5

Nanotech, Medicine, and the Pharmaceutical Industry

> Apparently, all our hopes about the 21st century medicine (nanomedicine) rely on using Z-DNA. One never knows.
>
> —*a reviewer of Robert Freitas' book* Nanomedicine, Volume I, *on Amazon.com*

> A highly readable exploration of a field that will play an important role in the evolution of our species.
>
> —*another reviewer of the same book, also on Amazon.com*

A Nanotech Paradox

As I write these words, the first baby boomers are about to turn sixty and, according to a cover article in *BusinessWeek* are both in search of ways to make them themselves look and feel younger and are peculiarly open to technological solutions to just about anything. Nano-enabled solutions seem like a way to enable their goals through new types of regenerative medicine and better drug delivery. The pharmaceutical industry meanwhile is seeing a basic challenge to their core business model, which is based in large measure on the concept of the "blockbuster drug," and these are apparently harder to find than they once were. Nanotechnology promises better drug discovery methods which could lead to new blockbuster drugs and through reinventing older blockbuster drugs with new delivery methodologies.

Given these technological, business, and demographic trends, you might think that nanotechnology would be having an immediate impact on the way that medicine and general healthcare is practiced and the way that the pharmaceutical business goes about its business. This does not seem to be the case, however. It is true that there is a considerable amount of attention being given to nanotechnology by researchers at important centers such as the National Institutes of Health in Maryland. There have also been some very early applications of nanomaterials for antibacterial purposes, specifically by Nucryst Pharmaceuticals and AcryMed. However, NanoMarkets' research indicates that at the commercial level, product managers in the relevant departments at pharmaceutical firms are fairly skeptical about what nanotechnology can bring to the table that will help them improve the bottom line for their companies. This has been confirmed by other market research firms and only the biggest nano-boosters, mainly either executives at the start-ups or science writer types, really seem to believe that there will be huge early markets in nano-enabled medical products.

This chapter explores where the opportunities actually lie in this field, both now and in the future. Before proceeding in this direction, though, I need to establish a couple of "ground rules" for what follows.

Timescapes

Perhaps the main reason why nanomedicine has not yet garnered much respect in either healthcare practice or the healthcare/pharma industry is that it appears to be a long way off and therefore will seem to many an area that is not worth more than a few casual thoughts at this point in time.

There are really two themes that are being intermingled here. The first has to with the relatively long periods that occur between the invention of a new drug or medical device. The other theme is the perception that nanomedicine is an essentially futuristic endeavor that is really not worth the time of serious physicians, businesspeople, or investors. These two aspects of the timeframe for the emergence of nanomedical and nanopharmaceutical products need to be separated out.

Regulation and Timeframes One of the main characteristics of the markets that we are examining in this chapter is extended times to market. Drugs, medical devices, and other related products must go through extensive testing both as a result of government mandates and because of concerns by the manufacturers themselves. It can be as long as decade between the time when a compound first shows some promise for curing or alleviating a disease and when it is generally available by prescription or over the counter, although the times to market for medical devices and for many drugs may be less than this.[78] However, because human lives are at stake, the product testing period is inevitably going to be longer than in most other sectors; certainly than in the two other sectors, those of electronics/semiconductors and energy, that I discuss in depth in this book.

For many, nanotech is interesting, but just too far into the future. In this chapter, I hope to establish that this is not really the case for specific products and developments. In general terms, it is also important, I believe, for skeptics to recognize the potential for nanomedicine of all kinds, because it is precisely this potential that may speed up times to market. In recent decades, at least in the United States, we have seen successful movements to cut the red tape and thereby speed up the approvals process on government approvals on various drugs. The argument is that cancer and AIDS patients simply can't wait for drugs to move through a painfully slow regulatory process.

Unless some drug that is rushed to market turns out to have disastrous health consequences, it seems safe to say that the pressures to get the both emerging drugs and medical devices to market faster will increase in the next decade. An increasing number of these drugs and devices will be nano-enabled. I would confidently expect that an aging boomer generation will certainly put pressure on the powers that be to make sure that that this happens, especially if the potential of the new nano-enabled technology appears to be significant, perhaps by adding years to life or by increasing cancer survival rates.

The "Fantastic Voyage" Syndrome The task of impacting public policy to ensure the rapid deployment of nanomedicine in sensible markets will no doubt fall to a few pioneer firms. Likely, there will be many skeptical laggards who will be left behind in way that significantly impacts their bottom line. One is reminded here of what happened to Kodak, which recognized that digital photography would eventually eclipse its analog (film) photographic technology and even built into its strategic plan some fairly optimistic growth projections for digital photography, yet still managed to underestimate the impact of digital photography on its traditional business.

As with Kodak, this may happen with certain medical devices and pharmaceutical firms.

However, if some firms aren't taking nanomedicine seriously enough, it may be because some individuals are taking nanomedicine way too seriously.

Because nanotechnology implies detailed engineering capabilities at about the same size scale as the cell, it is possible to dream up all sorts of elaborate medical procedures that can save people's lives in impossible circumstances, reengineer them completely, or even grant them complete immortality. Whole communities have grown up around such futuristic nano-enabled possibilities. Since it appears quite likely that much of what they are discussing will actually come about in the next couple of decades, I would not wish to criticize them too much. However, much of what they say sounds so futuristic that no business-person or investor with the usual one- to five-year perspective is going see profits in this futuristic version of nanomedicine and is consequently likely to brand all nanomedicine as a big "science" project.

I am going to call the tendency to get a little too excited about nanomedicine the "Fantastic Voyage" syndrome, after the Isaac Asimov story in which a submarine and its entire crew is miniaturized and then dispatched through the blood stream of a critically ill patient in order to fix a damaged heart. None of this had much to do with nanotechnology as the book and film appeared long before the term "nanotechnology" emerged. However, the idea that the miniaturization of complex medical machinery to the cell-sized level is implicit in most of the more futuristic visions of where nanomedicine is headed. As evidence to the influence of Asimov's story on current thinking in the circles supporting such visions, consider the title of the book on the "science behind radical life extension," by Ray Kurzweil and Terry Grossman: *Fantastic Voyage: Live Long Enough to Live Forever.*[79]

It turns out that the authors of this book mean the subtitle to be taken seriously. In the closing paragraphs of this chapter, I will discuss briefly what credence the practical businessperson or investor should give to such things.

The Tipping Point

Demographics and the potential technological power of nanomedicine seem to point to the emergence of a huge market for nanomedical products and drugs in the future. After all, the populations of developed nations are aging and nanomedicine operates at the cellular size level, which seems to suggest that it could be more effective than other forms of medicine. So if, as yet, there does not seem to be a groundswell of actual interest in the medical and pharma communities about such technology then some explanation is clearly needed of this apparent contradiction.

One possibility is that those of us who think that nanomedicine is a wave of the future are simply wrong. It would be arrogant of me to dismiss this possibility completely, but I will leave it to others to criticize what is, after all, a fundamental assumption of this chapter and (arguably) this entire book. Another possibility is that the apparent lack of interest in nanomedicine reflects the fact that it does not hold out an immediate promise of large revenues for the big pharmaceutical and medical device companies. As Clayton Christensen, the man who gave precise meaning to the term "disruptive technology" has pointed out,[80] the most revolutionary technologies are often ignored by large firms because they don't fit with their revenue expectations and business models.

Whatever the reason for the medical and pharmaceutical industries' nano-shyness at the present time, if my analysis of the emerging potential for nanomedicine is correct there will have to be some "tipping point," that will somehow make nanotechnology "respectable" in healthcare and pharmaceutical circles. Exactly what that will be is difficult to predict, but the most likely tipping points could be a well publicized success for some nanomedical technique or nanoengineered drug. This would either create a groundswell of interest in the medical community or among patients. Another possibility is that such techniques or drugs could be adopted by a prominent physician or (more likely) a well-known clinic. This would add credibility and drive the market forward.

Seven Ways Nanoengineering Will Contribute to Nano-Enabled Drug Discovery

One area where nanotechnology seems most likely to have a short-term impact is in drug discovery. NanoMarkets research indicates that the pharmaceutical industry faces increasingly challenging market conditions that are leading to an intensified search for better drug discovery technologies. Nanotechnology can help with that search.

The pharmaceutical industry has to discover and develop innovative medicines for a wide range of diseases in a marketplace that is both likely to experience growing regulatory and pricing pressures and that is increasingly targeted towards diseases that have been traditionally resistant to pharmacotherapy (e.g., solid tumors). In addition, as I have noted earlier, "big pharma" has built its business model around the discovery of a stream of blockbuster drugs that do not seem to be appearing with the regularity that this business model requires.

Today, nearly all pharmaceutical companies follow common technology processes for discovering drugs. These include cloning and expressing human receptors and enzymes in formats that allow high-throughput, automated screening (biochemical analysis) and the application of combinatorial chemistry. Thus, random screening can now be achieved with libraries sufficiently large and diverse to have a relatively high probability of finding a novel molecule. These libraries are possible because they can be generated by the techniques of combinatorial chemistry (combichem). Backing this are important developments in the application of genetics and genomics to understand associations between diseases and gene products. Importantly, bioinformatics are beginning to identify putative targets for a number of diseases.

An important consequence of today's approach is that more lead molecules are being discovered for diverse targets, giving the medicinal chemist more scope to find a candidate molecule. Yet the probability of success of launching a candidate molecule into the market remains basically unchanged at around 8 percent, and the perception is that the time to market is not reducing as might be expected with greater automation.

Although barely acknowledged outside a few specialist circles, it is becoming clear that many of the important solutions to the woes of the drug discovery sector will be nanotechnological in nature. This is because the size range that holds the most interest in drug discovery is more or less the same as that which defines nanotechnology from 100 nm down to the atomic level (approximately 0.2 nm). Nanotechnology cannot provide the ultimate solutions to all of drug discovery's problems. However, it is becoming clear that many of the tools developed to pursue nanotechnology and nanoscience will have important roles in drug discovery.

Atomic Force Microscopy While AFM microscopes (see Chapter 2) are invaluable for imaging objects at the nanoscale, until recently they haven't been able to see how components of a cell react in biological processes, such as their response to a specific chemical or compound. New imaging techniques involve attaching antibodies specific to individual proteins to the tip of an atomic force microscopes' probe. When an antibody reacts with the protein to which it is targeted, it creates a variance in the microscope's reading compared to a reading with a bare tip, showing the presence of a protein or other specific material in the region being scanned.

This technique leads to a greater understanding of the chemical dynamics involved in how cells react to stimuli, and could prove particularly significant for drug discovery. However, although the AFM is a useful tool to improve the understanding of molecular interactions, the widespread use of this technology in bioanalytical applications has been hindered by the limited throughput of these techniques and the high experimental burden imposed by complicated and expensive instrumentation. Highly parallel micrometer and submicrometer cantilever arrays, which are currently being developed, might increase the throughput of AFM-based force spectroscopy.

Near-Field Scanning Optical Microscopy (NSOM) allows the study of optical properties on the sample surface with a resolution better than the wavelength of the light. By scanning the optical probe at very small distances from the sample (a few nanometers), "evanescent waves" from the surface are detected by the probe. The use of evanescent waves allows bypassing of the wavelength limitation of traditional optical techniques (the traditional limit of resolution being half the wavelength in use).

NSOM can be used to image biological samples and could ultimately be a significant contributor to drug discovery. However, there is an obstacle to such deployment at the present time: manufacturing the probes reliably, the requirement for a very small hole through which the light passes being problematic.

Surface Plasmon Resonance SPR is a phenomenon that occurs when light is reflected off thin metal films and a small amount interacts with electrons in the film, reducing the light intensity. The refractive index of the materials sandwiching the film dictates the angle at which the light reduction (essentially a shadow) occurs. Using an SPR-based approach, the interaction of biomolecules can be detected in real time, offering applications largely for observing biological systems in action on a very small scale.

There is still a lot to be learned in this field, but NanoMarkets believes that there will be important improvements for imaging systems (and thus for drug discovery) as this learning process proceeds.

Nano Mass Spectrometry Current technologies used for proteomic studies are based on a variety of separation techniques followed by identification of the separated proteins and proteolytic peptides using mass spectrometry (MS). One popular technique is high-resolution two-dimensional (2D) gel electrophoresis, which is capable of resolving 2,500 or more distinct protein spots from complex samples. The in-gel protein spots are then identified using sensitive MS and sequence database searching. An alternative approach to 2D gel electrophoresis is chromatographic separation of peptides with electrospray ionization (ESI)-MS or tandem MS (MS/MS) detection.

The need for relatively large sample volumes is a major challenge for conventional ESI-MS analysis. There are now systems that address this challenge by applying nanotechnology to the electrospray nozzles. Thus, Advion's ESI Chip contains an array of nanoelectrospray nozzles, each one-fifth the diameter of a human hair, etched in a silicon wafer.

Dip-Pen Nanolithography I introduced DPN earlier in this book as one of the emerging nanotools that are making practical nanotechnology possible. In the context of drug discovery DPN serves as a scanning probe nanopatterning technique in which an AFM tip is used to deliver molecules to a surface via a solvent meniscus, which naturally forms in the ambient atmosphere. This direct-write technique offers high-resolution patterning capabilities for a number of molecular and biomolecular "inks" on a variety of substrates, such as metals, semiconductors, and monolayer functionalized surfaces.

DPN is becoming a significant tool for the scientist interested in fabricating and studying soft- and hard-matter on the nanoscale. DPN enables precise multiple patterns with near-perfect registration. It's both a fabrication and imaging tool, as the patterned areas can be imaged with clean or ink-coated tips. The ability to achieve precise alignment of multiple patterns is an additional advantage earned by using an AFM tip to write, as well as read, nanoscopic features on a surface. Taken together then, DPN makes a valuable tool for studying fundamental issues on colloid chemistry, surface science, and nanobiotechnology.

Nanoarrays The central paradigm in proteomics studies has been to identify differential protein levels in healthy and diseased cells, characterize these proteins and determine the protein's role in biochemical pathways. These proteins can then serve as diagnostic markers and potential drug targets.

Traditionally, 2D gel electrophoresis has been the workhorse in proteomics on the front end of mass spectroscopy. This technology, however, has been described as a bottleneck in high-throughput proteomic studies. As a result there has been a significant push to develop alternative technologies that serve to alleviate this rate failing step in protein separation prior to further analysis. At the forefront of these emerging and growing technologies are protein and DNA microarrays that allow the highly specific capture and analysis of a large number of proteins expressed in various cell types exposed to given perturbations in a high throughput manner.

Microarrays are generally defined as a substrate and surface chemistry onto which a biomolecule or capture agent has been immobilized for the purpose of expression analysis and functional characteristic. The potential to decrease the time required for the critical steps or protein separation and characterization are serving as a considerable driver in the protein microarray market. Microarray technologies can be considered platforms for nanoscale bioanalysis, and these products have already proven their value in the marketplace. However, currently available microarray technologies suffer from certain limitations that prohibit the exploitation of the full range of drug discovery applications. These limitations are already being addressed at the nano level with nanoarrays, which are ultraminiaturized versions of the traditional microarray that can measure interactions between individual molecules down to resolutions of as little as one nanometer. Nanoarrays are being touted as the next evolutionary step in the miniaturization of bioaffinity tests for proteins, nucleic acids, and receptor- ligand pairs.

Nanometer-scale resolution capabilities of nanoarray technology offer many advantages in the emerging field of functional proteomics. Unlike nucleic acids, which can easily be multiplied by amplification methods, individual proteins cannot be easily increased in quantity. Using nanoarray technology, very small quantities of individual proteins can be effectively screened against a large set of drug targets. In addition, nanoarrays can be incorporated as sensors in ways that would be impossible with larger microarrays. In addition, unlike traditional microarrays, a nanoarray can be used "in solution."

Quantum Dots As we discussed in an earlier chapter, quantum dots are nanometer-sized semiconductor crystals or electrostatically confined electrons. While other bioimaging tools such as fluorescent organic dyes have been available for many years, quantum dots hold the promise of a true revolution in bio-imaging quality and sensitivity.

Since 1914, researchers studying everything from blood tests to the effects of cancer drugs on the workings of a cell have used fluorescent dyes to tag cells. The dyes, however, can be problematic. Each dye molecule requires a laser of the same wavelength to cause it to illuminate. The dyes are also imprecise and have a tendency to bleed together. They can only be lit up for very short periods of time, typically just a few seconds after a light source is applied. Quantum dots take advantage of the quantum confinement effect predicted by quantum mechanical theory to fluoresce extremely brightly when excited by a light source such as a laser. They also don't have any of the disadvantages of conventional technologies. By varying the size of the crystals one can cause a rainbow of colors to fluoresce. In addition, quantum dots stay lit for much longer periods of time than conventional dyes, often hours or days.

Quantum dots enable the tagging of a variety of different biological components like proteins or different strands of DNA with specific colors. Quantum dots are designed to bond with and illuminate individual biological targets of choice whether genes, nucleic acids, proteins, small molecules, cancer cells, or even entire blood vessels. Some of the world's largest pharmaceutical and biotech companies, such as GlaxoSmithKline, Pfizer, and AstraZeneca, and a leading biotech firm, Genentech, are applying quantum dots in high-content drug screening and have completed initial drug screens using quantum dots as the biological readout. Genentech is applying quantum dots to the detection of breast cancer.

Six Ways That Nanotechnology Will Contribute to Drug Delivery

As the drug industry strives to meet the increasingly difficult task of producing new drugs, and especially new blockbuster drugs, nano-enabled drug discovery technologies that can improve R&D success rates and time to market will lead to substantial new business revenues over the next few years. Nanotechnology also promises some major opportunities in improved nano-enabled drug delivery systems. From a business perspective, these are important not only in providing better, more effective, better targeted, more profitable and less toxic drug delivery, but as a way of increasing/stretching the value of patents, since new products can be created from older drugs with new delivery systems.

The main opportunities for that are emerging for nano-enabled drug delivery systems are mostly based on delivering drugs in nanoparticulate form. Most drugs perform better as nanoparticles, because they can be targeted better and there are fewer side effects. For example, the anticancer drug, Abraxane, is essentially the same drug as Taxol (the active ingredient is paclitaxel). However the reformulation using nanotechnology overcomes some of the targeting and toxicity issues found in Taxol.

In addition, by using smaller amounts of drugs, treatment costs may be reduced. For example, according to NanoMarkets' research patients show a strong preference for nanoparticle inhalers as an alternative to the widely used injectable methods, which may permit far lower doses of expensive protein-based drugs like insulin. There are six types of drug delivery systems in which nanotechnology is likely to have an impact.

Injectable Delivery Systems Although injectable delivery systems have some inherent discomfort for the patient and other types of delivery may be preferred by both the patient and the physician, nanotechnology may be applied to produce improvements and additional revenue streams for the drug company. A case in point is Johnson & Johnson, which, at the beginning of 2005, announced that its Élan NanoCrystal technology would be used in a Phase III clinical trial for an injectable formulation of paliperidone palmitate, an antischizophrenia drug. This new "nano formulation" of an older drug which overcomes the original's insolubility, by reducing the particle size to under 200 nm. This new formulation enables the product life cycle to be extended by a significant amount.

Implantable Delivery Systems Implantable drug delivery systems are often preferable to the use of injectable drugs, because the latter frequently display first-order kinetics (e.g., initial concentration is elevated, which then drops exponentially over time). This may cause difficulties with toxicity, if and when the peak concentration remains above the therapeutic range. Conversely, drug efficacy often diminishes as the drug concentration falls below the targeted range. By contrast, implantable "time release" systems may help minimize peak plasma levels and reduce the risk of adverse reactions, allow for more predictable and extended duration of action, reduce the frequency of redosing, and improve patient acceptance and compliance.

To these advantages nanotechnology can add other benefits. Implantable drug delivery systems are expected to be used for the delivery of proteins, hormones, pain medications, and other pharmaceutical compounds. An example of how nanotechnology is already making its impact felt in this part of the drug delivery business is provided by pSivida's BioSilicon product. BioSilicon is a novel nanostructured material that effectively stores an active compound in nanosized pockets that controllably release minute amounts of drug as the silicon dissolves. pSivida is currently exploring biodegradable implantable methods for tissue engineering and ophthalmic delivery.

Among the first nanoscale devices to show promise in anticancer therapeutics and drug delivery are structures called "nanoshells," which may afford a degree of control never before seen in implantable drug delivery products. These nanoshells typically have a silicon core that is sealed in an outer metallic core. By manipulating the ratio of wall to core, the shells can be precisely tuned to scatter or absorb very specific wavelengths of light. For example, gold encased nanoshells have been used to convert light into heat, enabling the destruction of tumors by selective binding to malignant cells. A physician can use infrared rays to pass harmlessly through soft tissue, while initiating a lethal application of heat when the nanoshells are excited. Some researchers are experimenting with temperature-sensitive drug delivery control methods, using nanoshells that release their payload only when illuminated with the proper infrared wavelength.

Oral Delivery Systems The vast majority of consumers prefer an oral drug delivery system over one delivered intravenously. For this reason, R&D organizations are both seeking and finding ways to incorporate nanotechnology into traditional oral formulations. The challenge here for both traditional and nanomaterial-based delivery is to build systems that can survive the harsh environment of the human digestive tract. For example, researchers at the University of Texas at Austin described a means of using nanospheres for oral drug delivery. These nanosphere carriers are derived from hydrogels, which are highly stable organic compounds that swell when their environment becomes more acidic. They have been successfully formulated into controlled-release tablets and capsules, which release active compounds when the hydrogel body swells.

Nanomaterial-enhanced drugs offer increased oral bioavailability, and in some cases reducing undesirable side effects. Because of the blood brain barrier (BBB) most new chemical entities aimed at treating brain disorders are not clinically useful, but nanoparticles have been demonstrated to cross the BBB with little difficulty. Companies such as Germany's NanoPharm have developed systems capable of reaching the brain for anesthesia (Dalargin, an analgesic), cancer drugs, and various therapeutics. The company claims several advantages over existing systems, including (1) no requirement to open the BBB, (2) the ability to use potentially any drug, whether hydrophilic or hydrophobic, and (3) the drug does not need to be modified. The mechanism of action is not completely understood at this time. Another example is NanoDel Technologies of Germany, which uses polymeric nanoparticles that have drugs attached to their surface. These particles serve as "Trojan horses" for a wide variety of new chemical entities. Preclinical proof of concept was achieved with peptides such as tubocurarine, kyotorphin, loperamide, dalarkin, and cytostatic agents such as doxorubicin. The process also works with genes to achieve gene transfer, analgesia, or brain tumor therapy. By increasing bioavailability, nanoparticles can increase the yield in drug development and more importantly may help treat previously untreatable conditions.

One final note about the BBB. Although one of the unique advantages that can be offered by "nanodrugs" is their ability to cross the BBB, it is also one the aspects of nanodrugs that worry a lot of people, since it brings up the possibility of brain damage or psychoactive effects stemming from the use of drugs whose main objective has nothing to do with the brain.

Topical Delivery Systems Nanomaterials also provide a unique opportunity for rapid topical delivery of active compounds. Given their very small size, nanomaterials are able to enter human tissues and cells quickly. Companies such as Novavax have developed regulated topical systems that take advantage of the unique properties of micellar nanoparticles. They have developed two hormone replacement therapies dubbed Estrasorb (which received FDA approval in October 2003) and Androsorb (which successfully completed Phase I human trials in 2003).

Toxin Removal Colloidal dispersions have been demonstrated to remove potentially lethal compounds from the bloodstream, including high concentrations of lipophilic therapeutics, illegal drugs, and chemical and biological agents. A team of scientists at the University of Florida and Clarkson University in Potsdam, New York have demonstrated favorable results to this end, using biocompatible microemulsions. These oil-in-water systems have a rapid and efficient absorption capacity for many target molecules that are frequently overdosed, albeit intentional or accidental. The microemulsions use a polymeric surfactant, in combination with an ionic co-surfactant. Other oil compositions have been made with varying absorption characteristics for different toxins, in all cases avoiding hemolysis and unwanted blood coagulation. The mechanism by which nanoparticles trap toxins involves aromatic or cyclodextrin receptors with electron deficiencies, which form complexes with selected toxins, sometimes referred to as "benzenoid moieties."

Transdermal Systems The number of FDA-approved polymers available for use on skin or medical use is increasing rapidly. With this in mind, the industry will be presented with opportunities to create new transdermal platform designs that improve on-skin properties and diffusion of active molecules from patches. This trend is expected to result in smaller and less invasive patches that increase the universe of available drug candidates. In some cases, electronics are even being integrated into patch-like platforms involving wound care, monitoring, and diagnostic methods. As such, patch-like platforms will play an important role in healthcare to treat, measure, diagnose, and generally improve quality of life.

Nanotechnology and Medical Diagnostics

I have already discussed the use of nanotechnology to build nanobiosensors in a previous chapter. These devices have a number of important applications including homeland security. They may also be used to improve medical diagnostics at various levels.

The most obvious way in which nanotechnology can help with medical diagnostics is through the building of improved labs-on-a-chip. Such labs on a chip exist now and are based on microtechnology, especially microfluidics. While they are obviously something less than entire labs miniaturized in the tradition of *Fantastic Voyage*, they have increased the efficiency of diagnosis by reducing the need for valuable real estate and technical staffing and by speeding up the diagnostic process. With the cost of medical care increasing fast in many countries and given how early diagnosis can affect the ability of patients to survive a disease, no further justification for labs-on-chip need be given.

Nanotechnology's impact on labs-on-chip will not be revolutionary, but will rather make these devices more sensitive. For example, it has been demonstrated that a nanosensor using nanowires can detect a single virus. Nanotube based sensors would be sensitive enough to probe the structure of very small samples of DNA and proteins. This type of capability will certainly have important R&D applications and probably practical medical ones too, although exactly what these are remains to be seen. Nanoengineered labs-on-a-chip will no doubt improve and become more distinguished from microengineered labs-on-a-chip in terms of performance as nanofluidics becomes more developed.

Whatever the future of nanoengineered labs-on-a-chip, they are still essentially a nanoelectronics application and much of what I have said about nanoelectronics applies to them. However, we are talking about the opposite of mass market chips here, which better enables some of the more advanced nanoengineering tools to be brought into play to create them. In particular, both dip-pen nanolithography and nanoimprint lithography seem to have a considerable potential for creating the complex nanostructures that will underpin the nanoengineered lab-on-a-chip of the future. It is possible that lessons learned on this application could help improve the performance and manufacturing platforms for nanosensors and for nanoelectronics more generally.

In a similar way to how nanotechnology has the potential to take labs-on-a-chip to the next level, it also has the facility to improve medical imaging. While labs-on-chip provide data for diagnostics performed outside the body, medical imaging provides diagnostics from within the body, typically using X-rays, CT scanning, MRI, and ultrasound. In addition to making primary diagnoses these imaging modalities can also be used to monitor the progress of diseases and the impact of drugs and other therapies.

Where nanotech fits into the picture is primarily in the area of molecular imaging. This is an application that is even more like the miniaturized submarine of *Fantastic Voyage*, but in this case the "sub" is usually a nanoengineered "smart probe," which can identify diseased tissue, as for example, tumors, and transport quantitative and qualitative information back to a computer monitor. Again molecular imaging of this kind is really just an extension of older imaging modalities that used dyes injected into the bloodstream to show up abnormalities and again molecular imaging of this kind is really just a very specialized form of nanosensor.

The hope is that nano-enabled molecular imaging will, like labs-on-a-chip, diagnose diseases earlier than would otherwise be the case and also that this kind of imaging could be less invasive than current procedures. This would certainly be the case if molecular imaging proves to be a replacement for biopsies in certain instances. From the perspective of opportunities in the nanotechnology sector, three nanomaterials platforms that seem like they will have a significant impact on imaging are quantum dendrimers, buckyballs, quantum dots, and carbon nanotubes.

Dendrimers These are a nanostructures that we have not discussed before. Basically they are a polymer-based material with a treelike structure and they have various potential nanobiological and sensing applications. In the context of imaging, they have been used in a variety of ways, including making cells fluoresce when they are disturbed in certain ways, thereby enabling physicians to tell if drugs are doing their job or if viruses are attacking the cell, and so forth.

Buckyballs Buckyballs (or Fullerenes) also have the potential to be used in medical imaging, but in this case they are used in conjunction with contrast agents. Contrast agents are chemicals that improve the efficacy of imaging by increasing the image resolution and brightness of the image being sent from the site under investigation. Basically, the idea here is the contrast agent is house in the buckyball nanostructure. The objective of doing this is mainly to reduce the toxicity of the contrast agents itself and to increase their power.

The particular kind of fullerene being used in this application is a "metafullerene," which is a few metal atoms encased within a fullerene. (A specific example would be a gadolinium metafullerene—gadolinium being one of the most widely used contrast agents.) The "encasing" prevents the contrast agent from traveling too far into the body and causing toxic effects. This is not a huge issue with gadolinium (although it *is* an issue), but it can be a very important factor when radioactive materials are used as contrast agents, which they frequently are. In terms of increasing the power of contrast agents, under certain circumstances, the buckyballs seem to have something of a catalytic effect. For example, a gadolinium metafullerene may be engineered in such a way that it is a more powerful contrast agent than gadolinium alone.

Quantum Dots We have already met the idea of using quantum dots to make cells fluoresce in the context and this concept can be easily extended to diagnostic medicine. Researchers have already shown that quantum dots can be a useful diagnostic tool for identifying cancers of various types. Researchers at Emory University and Georgia Tech in collaboration with Cambridge Research and Instrumentation have used cadmium selenide-zinc sulphide quantum dots to identify tumors in mice, for example.

Quantum dots' potential role in nanomedicine appears to be very significant. Not only do they have a part to play in drug discovery and diagnostics, but they may also serve in therapeutic modalities, where their large surface areas make them well suited for certain kinds of drug delivery. As such they are attracting attention from some major firms such as Matsushita and Sumitomo Biosciences and Invitrogen (which acquired Quantum Dot Corporation). There are a number of areas for research and sources for competitive advantage in this area including the specific chemistries for the quantum dots, coatings for the quantum dots to reduce toxicity, and improved delivery methods.

Carbon Nanotubes Among the many fantastic properties of carbon nanotubes is an ability to fluoresce in the near infrared part of the spectrum and may find some biomedical imaging applications as a result. Some research has been done in this area at both Rice University and the University of Illinois, but this application for carbon nanotubes seems to be at an earlier stage of commercial development for medical imaging than fullerenes, dendrimers, or quantum dots. And there are certainly fewer parties interested in CNT-based imaging than in the other materials platforms mentioned above.

Nonetheless, I believe that CNTs may present some important opportunities in medical imaging in the next five to ten years. This is because, more so than for the other nanomaterials platforms mentioned in this chapter, CNTs are likely to become a widely available and standardized material in the coming decade In addition to the likely high level of availability of carbon nanotubes, CNTs have the important advantage that they are nontoxic and that the wavelengths at which they fluoresce can be tuned by adjusting the physical dimensions of the nanotubes. This means that more complex information can be imaged and conveyed to the physician or researcher. All this opens the way to using CNTs in a similar way to the quantum dot technology described above and for a broad range of diseases. As an example, at the University of Illinois researchers tracked glucose in diabetics.

Nanotechnology and Regenerative Medicine

I began this chapter by discussing the fact that the aging of the population in the Western world and in Japan would be a key driver for new medical technologies and particularly for nanomedicine. This is most obviously the case in the area of regenerative medicine, the restoring or replacing damaged tissues, bones, and organs. Some of this regenerative medicine is already available, while others are coming soon. Some of it is highly futuristic. (More on that aspect of nanomedicine at the end of this chapter.) Some of it is the nano-equivalent of cosmetic surgery. Some of it is miracle medicine for the severely injured and disabled.

Nanomaterials may actually be better, in that they are stronger and more durable than the real thing. A burn victim may have his or her burned skin replaced with a nanomaterial that looks and feels like real skin, but is much more durable than real skin. However, this durability may translate into people who have undergone the procedure looking much younger than their actual chronological age and it is not unlikely that in such a case new cosmetic surgeries will emerge that use the same technology to make 60-year-olds look 40 years old. It is possible that the technology could be used as a step to making lifelike robots for both work and pleasure applications.

All this raises ethical and societal issues that really go well beyond the scope of the book, although I will discuss the question of how seriously the businessperson should take "transhumanism," at the end of this chapter. For the time being, I want to summarize a few of the ways that nanotechnology is showing some promise for regenerative medicine for relatively near-term applications, likely within the next decade. It is probably fair to say, however, that while the pharmaceutical and diagnostic applications outlined above are pretty much with us now, nanoengineered regenerative medicine in any of its likely forms is still some way off.

The "unique selling proposition," of nanotechnology in the context of regenerative medicine is that it can be used to build "spare parts" for the body that are finely enough sculptured and textured to make them excellent substitutes for the real thing.

Nanogels One of the nearest to commercialization applications in regenerative medicine for nanomaterials are gels that provide structures—much like trellises used to "train" plants—over which damaged nerve (and other) cells can grow as they regenerate. Similar gels have been on the market for a while, which should help the nanogels find acceptability quickly. The advantage that nanoengineering brings to the table here is that a nanoscale trellis fine-tunes the regenerative process, so that much of the original functionality is regained.

Organ Replacements Completely artificial organs have been around for several decades, but have never really become as popular as once thought. The biggest example of this is the artificial heart. Meanwhile, there is a long wait for transplants of human organs and the procedures for making these transplants are lengthy, expensive, and dangerous.

It is possible that new nanomaterials, coupled with nanoelectronic devices could make very significant contributions to constructing artificial organs that would go some way at least to make them a practical alternative to human transplants and thereby alleviate the shortage of organ donors. A more interesting, and perhaps, even a more likely direction, is to use "nanotrellises" of the kind described above to grow complete organs. According to one report, NASA has used this approach to grow heart cells and connect them up in a way that actually allows them to "beat" when put in the correct artificial environment.

Growing a complete heart is a very long way off and may seem more difficult to achieve than constructing a completely artificial heart. However, this is not necessarily the case, since in a "heart growing factory" all that needs to be done is to provide structure for the heart cells to correctly self-assemble. Creating an entirely new heart from scratch means matching numerous performance parameters, which may be easier said than done. Also, it is fair to say that while the heart offers the most dramatic example of this kind of procedure, the likelihood is that the commercialization of this kind of technology will begin with the skin, as it is the body's largest organ. It is skin growth that is the nearer-term business opportunity here.

Better Blood A number of other less dramatic nano-enabled procedures should also help improve the cardiovascular system. There has been talk of creating artificial blood cells, which would consist of nanospheres filled with high-pressure oxygen that could be injected into the body. This would be as much as a drug delivery system and could be used to help treat heart attack or stroke victims. Or it could be used to enhance performance, perhaps becoming the next big business opportunities after oxygen bars? Incidentally, artificial blood cells already have a name: respirocytes.

A somewhat less futuristic approach to dealing with cardiovascular problems is what we shall dub a "nanostent." A stent is a device intended to keep clogged vessels open and they have been around for a while. Nanoengineering, however, is being applied by at least one firm, Advanced Bio Prosthetic Surfaces (ABPS), to make stents better. In this case, a nanoporous coating is being used to ensure that the stent has no rough edges that could hurt the blood vessel or cause inflammation and it also ensure that the stent is stronger and more flexible.

Improved Memory Primarily aimed at Alzheimer's patients, there are now several nano-related efforts to improve memory. These could also ultimately find a way into improving memory for the general population, if they proved to be effective and not too expensive.

Much like the heart example above the choice is between a completely artificial solution using some kind of computer memory (nanomemory or not) and a nanoengineered conductive polymer to make the connection to the brain. Another possibility would be to take real neurons and plant them in the brain using some kind of nanoengineered vehicle based (perhaps) on some of the drug delivery systems that we have already met. In this particular case, the nanopackaging for the neuron would have to be such that the neuron could be artificially stimulated by a source (perhaps a nanobattery) in order to enable it to emit chemical neurotransmitters.

Timeframes, Futuristic Nanomedicine, and a Business Perspective on "Transhumanism"

Most of the nanomedicine and nanoengineered drug technology that I have discussed in this chapter so far is actually being used or is in the productization stage. In some cases, the nanotechnology involved is actually more evolutionary than anything else. For example, as I have already noted, labs-on-a-chip have been around for a while using microtechnology of various kinds. No one expects nanoengineered labs-on-a-chip to exactly transform medicine. Similarly, the nanoengineered gels that are showing promise as way of creating structures for healing cells are just the next stage in the evolution of similar gels that have been around for some time, albeit with a cruder deep structure, but which perform similar functions. In a few cases, such as that of growing hearts, the technology still seems a long way off.

With all this said and even considering the fact that the healthcare and big pharmaceutical industries do not seem to be overly enthusiastic about nanomedicine at the present time, I believe that nanomedicine will be where the biggest opportunities for nanotechnology will be found a decade from now. The potential for increased longevity that nanotechnology seems to present to us, coupled with the aging population seems to be an irresistible force, always assuming that nanomedicine can live up to its potential.

How far, then, should one take all this? Some proponents of nanomedicine show little interest in the relatively modest developments that are the focus of most of this chapter. Instead they are focused on radical improvements in longevity and actual immortality. People who advocate using nanotechnology in this way are often referred to as "transhumanists" and, as their name implies, they are people looking to technology to deliver an *ubermensch* that transcends homo sapiens in both its physical and mental characteristics. Nanomedicine, or even nanotechnology as a whole, is certainly not the only technology that is invoked by transhumanists. While computer science, biotechnology, and robotics are also in their armory, nanotechnology is a big (and growing) part of what they talk about.

There are various possible reactions to transhumanism. In an earlier chapter we looked briefly at the position that transhumanism is fundamentally immoral and is an idea that should be (at the very least) discouraged and suggested that an antinanotech movement may emerge that will have to be countered through lobbying and PR. At the beginning of this chapter, we suggested that the speculative, and potentially highly controversial nature, of nanomedicine is actually a factor that is putting some pharmaceutical and healthcare companies off from becoming involved with nanomedicine. Put crudely, I am suggesting that firms, as well as individuals for that matter, are not putting their money into nanomedicine because they perceive it as being just too weird!

Just how weird is transhumanist medicine really? While it may test our moral and religious sensibilities to the extreme, there can be no doubt that products and services that can guarantee extreme old age and even immortality would command a huge market, assuming that the price of these products and services is not too high.[81] A similar and more immediate set of issues is raised by human cloning, the implications of which appall many of us, but which, should it become available as a service, will no doubt find many ready customers.

It is therefore worth taking a brief look at whether the transhumanists claims have anything in them at merits serious attention by a businessperson; the alternatives being that it is something that is so far off in the future that the discounted current revenue from the transhumanist project is effectively zero or that this project is little more than self-delusion.

I suspect that a case can be made for any one of these possibilities. My personal view is that business planners, who have a time horizon that extends well into the next decade, should not be too dismissive of transhumanism. A full discussion of why I believe this would take another book at least as long as this and involve a discussion of developments in semiconductors, artificial intelligence, biomimicry, and biotechnology in addition to nanotechnology. However, if I were ever to write such a book, I would argue for something that I might call "weak transhumanism."

Let me explain. The central dogma of transhumanism is that human technologies grow in an exponential way. The two distinctive features of an exponential curve is (1) that despite the common idea that exponential growth is very rapid growth, exponential curves begin with a shallow incline, and (2) when the curve starts growing extremely fast this occurs very suddenly. Transhumanists take all this to indicate that we are moving towards a "singularity," a point at which our technology will make us more than human. Personally, I am not completely comfortable with this concept, although I would not dismiss the idea completely. I don't think that one needs to go all the way with the transhumanists to see some dramatic new business opportunities emerge from the type of technologies that transhumanists like to consider.

Even if the long-term exponential curve idea may turn out to be fanciful, it seems to me that there are really take off points after a long period of R&D has been completed. Many of the procedures discussed in this chapter seem to be at the point at which many of the really difficult problems have been solved and they are ready to go into clinical trials. Once some of these procedures are tried and tested on sick patients, how long will it be before they are used to make us more than human? I am thinking especially about regenerative medicine here.

Indeed, there would be nothing that new about this. As Andy Clark has indicated in his excellent book about cyborgs,[82] there is nothing really new about any of this. We have been doing this kind of thing ever since we began wearing spectacles. It should also be remembered that what I have presented here is really the tip of the iceberg. I have looked for and discussed applications that are clearly nanotechnology, but there are huge leaps forward in medicine promised by various biotechnologies and bioengineering techniques that would not exactly qualify as nanotechnology.

As to actual immortality, I am not brave or foolish enough to predict it. But imagine we could come up with a biotechnological technique to produce the enzyme telomerase and a nanoengineered delivery vehicle that would deliver it to the ends of aging chromosomes to reactivate cell division, we would have a dramatic rejuvenation process that would in effect be a reasonable approximation to immortality. We are not really anywhere near having a technique like this, but my point is that put in these rather materialistic terms we are no longer talking about immortality as a miracle (or a curse), but rather as a comprehensible medical procedure.

Summary: Key Takeaways from This Chapter

This chapter has primarily dealt with shorter-term opportunities and here are some of the key takeaways from it:

1. Its strongest advocates tell us that nanomedicine will create medical miracles. And so it may. But from the point of view of practical business considerations, however, it is important to separate out area where nanomedicine seems likely to produce revenues in the next few years and areas where nanomedicine has genuine potential for spectacular results as in enabling super-enhanced longevity, for example. At best, however, we are the stage where what is being investigated is basically the fundamental science, not productization, infrastructure or business issues.
2. Despite the obvious potential of nanomedicine, there are many in the healthcare industry and in big pharmaceutical firms who remain highly skeptical of whether nanotechnology is really going to take off in their areas. A careful look at demographics and the already-established capabilities of nanotechnology in medicine suggest that it will be hard for skeptics to retain their stance forever. All it will take will be a tipping point in the form of a nano-influenced procedure becoming popular or some influential person endorsing such a procedure.

3. Nanotechnology seems set to provide some major benefits to the pharmaceutical industry in the near-term future, both in terms of drug discovery and in terms of drug delivery. These will enable big pharmaceutical firms to find new blockbuster drugs and add to the life of those already on the market or in the pipeline.
4. Nanotechnology will also help improve diagnostics, but mostly in an evolutionary way, especially by offering ways to enhance labs-on-a-chip and medical imaging.
5. There are many ways in which nanotechnology can lead to a great leap forward for regenerative medicine. These include some very near-term opportunities, such as stronger, safer stents. There are also interesting possibilities for using nano-engineered "trellises" to help grow damaged cells.
6. Transhumanism is the idea that developments in nanotechnology, robotics, and biotechnology will eventually lead to us becoming more than human. There is a somewhat infantile tone to many of the pronouncements, but a closer look at some of their claims make them look a lot less silly. This direction for nanomedicine is not only fascinating, but also potentially profitable.

Further Reading

Unlike many of the individual application areas to which nanotechnology is being used or potentially could be used, nanomedicine is beginning to develop a rich and broad literature all its own. A search under the terms "nanomedicine" or "nanobiotechnology," on one of the major Web engines produces thousands of references.

Readers interested in pursuing the issue of nanomedicine further might consider beginning in a gentle fashion by reading Asimov's original *Fantastic Voyage*, or at least seeing the movie. A more serious student might begin by perusing the books of Robert A. Freitas, who is writing a series of books under the general heading "Nanomedicine." According to Amazon.com, at the present time, two volumes are currently available; *Basic Capabilities* and *Biocompatibility*.[83] These come with the warning that, although very thought-provoking, they come from the more Drexlerian wing of the nanotech sector.

There are also a couple of books on the interface between the nanotechnology and biotechnology, two topics that are really quite closely related and might have been thought of as a single topic but for their very different history. One of these books is *Bionanotechnology: Lessons from Nature*, by David Goodsell.[84] The other is *Nanobiotechnology: Concepts, Applications and Perspectives*, edited Christof M. Niemeyer and Chad A. Merkin.[85] Whether there is a difference between bionanotechnology and nanobiotechnology, I will leave the reader to decide for himself or herself.

Outside of such major texts, the Web sites of both the Foresight Institute and the National Institutes of Health (NIH) give significant coverage to nanomedical issues.

6

Spreading Nanotech: Industry-Specific Opportunities and Future Opportunities

...what lies before us are tiny matters...
—*Ralph Waldo Emerson*[86]

Any sufficiently advanced technology is indistinguishable from magic.
—*Arthur C. Clarke*

The Coming of the Nano Economy

As we have seen, much of both the commercial and R&D activity in nanotechnology falls into the area of nanomaterials or nanostructures of some kind. In terms of creating complex products, at a rough guess, about 80 percent of all nanotech activity is, and will be for some time, accounted for in some way by semiconductors, electronics, energy, and life-sciences-related sectors; that is, the "big three" sectors we have discussed in depth in the chapters earlier in this book.

However, the impact of nanotech will be broader than the previous three chapters read alone might suggest. Much of the claim to fame of nanotech is that its impact will be felt everywhere.[87] The impact of nanotechnology will often come through products created in the big three nanotech sectors. Nano-enabled mobile communications and energy use will improve the transportation sector, for example, making transportation more convenient safer, cheaper, and more pleasant to use. Nanosensors—which fall into the nanoelectronics sector—have myriad uses in the food and agriculture segment. And nano-enabled drug delivery systems may also improve agriculture through healthier livestock.

In the end, it is the particularities of the industry sectors that matter in terms of assessing the opportunities that nanotechnology presents in each sector. So rather than write an all-too-broad account of how nanotech will be everywhere and make the world a better place, I am using this chapter to show how complex nanotech products will change five very different parts of the economy: the chemical industry, textiles, the construction industry, the food/agriculture industry, and robotics. It would have been easy to pick other sectors that are also important and also where significant impact of nanotechnology is to be expected. However, it would have been impossible to cover all such sectors in detail in a book that is anything less than an encyclopedia, which this book decidedly is not. Of the five sectors that I have chosen to cover, four represent significant chunks of the world economy. Robotics, the other segment, could potentially alter the way that manufacturing is done around the world. All will be changed by nanotech, although in different ways. The chemical industry may be transformed into a nanomaterials industry. Nanotech may change the competitive factors in the textile industry, create new products for the building industry and make the food and agriculture industry more efficient. It may also bring robotics a little closer to the vision that most of us had when we were kids.

This chapter then examines the impact of nanotech on the areas of the economy listed above and provides hints of where the opportunities are likely to be found in them. As in previous chapters, I will try to work backwards from the major demand and market evolution patterns in the sector to the need for nanotech. As in all parts of the economy, nanotech will not make its impact felt unless it captures the imagination of customer and hence offers the potential to make money for suppliers. Finally, as in all of this book, I have tried to focus on relatively short term possibilities, such as might interest a practical businessperson. However, if any of the applications that I discuss below seem somewhat unlikely to reach commercialization, I can only refer you to the quotation from Clarke at the beginning of this chapter.

From Chemical Industry to Nanomaterials Industry

The chemicals industry is divided into two parts. The specialty chemicals industry is distinguished from the bulk chemicals industry by the fact that it sells by the bottle, gram, or pound and not by the ton or truckload. Specialty chemicals may be one of the first areas to see significant revenues emerge from nanotechnology. As we saw in a previous chapter, new nanomaterials are being developed at an ever-quickening pace and stripped of the high-tech "nanomaterials" name, these coatings, powders, gels, and so on are really no more that a particular type of specialty chemical.

Some of the biggest firms in the chemical industry, BASF and DuPont, for example, are already heavily committed to nanotech. The major firms in the specialty chemical sector are well positioned to take advantage of the commercialization of nanotubes as these products move from being a fascinating freak of nature to be worked on by researchers at major universities to becoming a major material that must be supplied in significant quantities.

As this process occurs, I would expect the established specialty chemical industry will take over from the some of the smaller firms that are now major suppliers of nanotubes. The economic reason for this is that carbon nanotubes are likely to become commoditized[88] (or nearly so) and the existing specialty chemicals and materials firms are already very good in bringing economies of scale in production and marketing to bear on this kind of product. (Whether the volumes involved ever put nanomaterials into the bulk chemicals business remains to be seen.)[89]

A similar story will be told for nanomaterials other than carbon nanotubes, although their extremely attractive characteristics give CNTs a special place in the nanotechnology pantheon that no other nanomaterial can make claim to. However, it is certainly the case that some important chemical and materials industry firms are paying a lot of attention to nanotechnology. For example, at a nanotechnology conference at which I spoke in 2004, a representative from DuPont stated publicly that it is rebuilding specifically to be in tune with the nanotech era. Many chemicals and materials firms are simply watching nanotech closely to see what happens, realizing that nanotech may be an opportunity or a threat to their existing core business. Sometimes the presence is deeper than may be obvious at first sight. While they may not be planning to remake their businesses with nanotech in mind, some important chemical and materials firms are strategic investors in nanotech startups or partners with these start-ups in some other manner. This is usually not front page news or even the subject of much comment in annual reports, that is, unless some exciting material arises as the result of the collaboration.

As the nanomaterials business begins to ramp up, I would expect to see a spate of acquisitions of smaller suppliers by some of the established chemical and materials firms anxious to get their fingers in the nanotech pie. But it should also be noted that a fewnanomaterials suppliers—I am thinking especially of carbon nanotube manufacturers—have already done a good job building their own brand names. This may be even more reason why they will be acquired by bigger firms with more marketing and production muscle. It also may be a reason why some specialist nanotube firms may be able to remain independent or do an IPO. It is sheer speculation, but it seems plausible that it would be a firm of this kind whose IPO is so successful that it touches off a nanotech boom in the way that Netscape's IPO set of the dot-com boom. If this were to happen, it may or may not be a good thing for nanotech.[90]

Eventually, nanomaterials may actually come to be thought of as an important sector of the specialty chemical industry. Indeed, it is just about possible to imagine a future in which nanomaterials have become so important to the world that "specialty chemicals," is a term that has ceased to be used that much and this sector is simply considered to be the nanomaterials sector, with some of the older specialty chemicals simply being replaced by nanopowdersor nanocoatings, or becoming of little importance to the industry sector as a whole. It is far from clear that this is the way that things will turn out or that they need to in order for important new opportunities to arise in the specialty chemicals sector specifically because of nanotechnology.

In Table 6.1 I have outlined how the chemical industry might evolve in an era in which nanotechnology becomes increasingly dominant in the industry's collective thoughts and in the world in general. Loosely speaking, something similar to this table could be drawn up for many of today's industries (see Chapter 7). But because nanomaterials are lower down on the development chain than more complex products, such as for example a nano-enabled drug delivery system, the kind of evolution described in Table 6.1 may well occur their first. If, as is possible, nanomaterials become the major growth segment of the specialty chemicals industry, then the boundaries between the nanotech sector and the specialty chemicals sector may become harder and harder to define, just as they have for the semiconductor industry that we discussed in depth in Chapter 3.

In the end there will be many ways in which the specialty chemical industry could participate, and find opportunities, from nanomaterials. We have indicated several strategies that firms in this industry might employ in Table 6.2, along with some current examples. The list provided in Table 6.1 almost certainly is not exhaustive.

Nanotextiles and "Intelligent Clothing"

While the semiconductor and chemicals industry may eventually become so permeated by nanotech that it will become hard to distinguish where the traditional industry ends and nanotech begins, I don't think that anything like this is

Table 6.1
The Evolution of a Nanomaterials Industry: A Prophesy

Stage	Description
Nanomaterials start-ups and corporate projects	Spinoffs from universities begin to proliferate with VC, university, and government funds. Some large chemical and materials firms also run internal projects to develop specific nanomaterials or provide some low-level funding for budding nanomaterials start-ups. Most larger firms just watch from afar. Focus is primarily on R&D, with only a passing consideration of commercialization. The period in which this kind of activity predominates could last up to five years.
More focus from big firms and Wall Street	Something sparks off a lot of interest in this sector—perhaps it will be a particular material that captures the public's imagination[91] and its dollars. Or perhaps it will be a highly successful IPO in this space. Investment banks and VCs start to take a lot more notice of the sector, although some of them will be too late into the sector. Much the same thing can also be said about large chemical firms and materials firms who also want to get into the nanomaterials boom. Acquisitions will begin to happen in significant numbers, as will IPOs of nanomaterials firms that have built some kind of brand name. This phase of the evolution of a nanomaterials industry may last from one to three years and may include a big nano boom.
Nanomaterials as big business	Revenues from novel nanomaterials begin to reach levels where they impact bottom lines on the financial statements of publicly held companies. Some of these firms may even set up nanomaterials divisions and a few of these divisions could ultimately be spun off as the independent companies, if the parent firms believe that the markets, risks and level of required entrepreneurialism and (perhaps) customer base are significantly different from their core business. This part of the evolution of a nanomaterials business may last several years, or even a decade, culminating in a relative maturity for the nanomaterials industry sector
The chemical industry becomes the nanomaterials industry	This phase may never really fully occur, unless nanoengineered materials become the norm throughout the chemical and materials industry, including the bulk chemicals sector. Imagine some nanomaterial replacing cement or sulfur, for example (it's not that easy to believe). However, in specialty chemicals it is possible that at some time in the future nanomaterials will come to predominate in much the same way that engineered drugs have replaced herbal remedies in "professional" medical circles.

Table 6.2
Specialty Chemical and Materials Industry Nanotech Strategies

Strategy	Comment	Example
Remake firm to take advantage of the coming nanotech era	Lots of risk in this strategy—the nanotech era may never come and rebranding is a difficult goal to achieve successfully. On the other hand, firms such as IBM have remade themselves in ways that an old, established chemical firm can only envy.	DuPont is taking this approach. It is not likely to have many emulators.
Become a supplier of materials for a user developed application.	Avoids high upfront R&D costs, which will be attractive to many players in a world in which firms would rather shift such costs to universities or the government.	Seagate has developed diamond hard coatings for its disk drives using buckyballs. If this approach proves important commercially, then Seagate is likely to want to buy such coatings from an outside supplier.
Use nanotech to add value to existing products	This is probably going to be the strategy that most firms in the chemicals industry adopt with regard to nanotech, using it to make more from their core business.	Both DuPont and Exxon are trying to develop stronger polymers by mixing or bonding in buckyballs.
Invest at a distance	Could involve direct investment in a nanotech start-up firm or investment through a VC arm.	BASF has a VC fund in which nanotech figures highly.
Spinoff	As noted in the main text, a large chemicals or materials firm may decide that the nature of a nanotech business is best handled through a smaller organization, to which it will give financial and sales support as needed.	3M spins off AVEKA (nano-particle processing) or Dow spins off Aveso (which makes tiny printable displays).

ever likely to occur in the textile industry. While relatively few people have an undying affection for a traditional specialty chemical, the same is certainly not true for fabrics. It's hard to imagine nanotechnology producing materials that would make people give up their desire for fine natural fabrics, namely, silks, wool, linen, and so on. Certainly, there have been many attempts to substitute man-made fabrics for these natural products, and they have succeeded up to a point, but *only* up to a point. However impressive the performance characteristics of "nanotextiles" turn out to be, they may never completely replace the "real thing."

That said, those performance characteristics really do seem to be quite impressive, and early to market. Indeed, one of the first things that a casual reader learns about nanotech is that one of its early applications is in the textile industry. The reference is usually to NanoTex, which is part of Burlington Fabrics. So far, NanoTex's most famous product is its spill-resistant material, which in the words of NanoTex expels liquids that would otherwise stain, like water off a duck's back. Other nanofabrics that have been promised by NanoTex include quick-drying cloth and a synthetic cotton. Beyond NanoTex other interesting textile related products are being developed in both industrial labs and universities and many of them have obvious commercial potential. For example, Clemson University has developed a self-cleaning fabric, based on a thin-film coating that uses silver nanoparticles and which is used to cover fabrics. The nanostructures in this coating are such that dirt literally bounces off when the fabric is washed.

The NanoTex and Clemson examples, illustrate a key direction for nanotextiles—cleaner clothes! Since, earlier in this book we have talked about such weighty matters as improved longevity, solutions to the supposed energy crisis and new directions in electronics, there is something almost humorous about talking about cleaner clothes. Yet the appeal of this kind of product hardly requires an explanation. Here are products that can sell in the market today against existing products over which they have significant performance advantages. Much the same thing can be said about the heat-resistant, lighter, and stronger fabrics that nanotechnology is also likely to create. Consider the issue of strength for a minute. It is easy to imagine nanotextiles created using carbon nanotube coatings of mixtures. This is probably overkill for most applications, but could be the next revolution in military and police wear—better than Kevlar for bulletproof vests.

An even more interesting direction for nanotextiles is the trend towards to "intelligent" fabrics. There are at least two dimensions in which a fabric can be intelligent. One of these is the "wearable computer," trend, in which computers are embedded in clothing to provide an always-on interface to powerful processors for the person wearing the clothing. These processors can help provide useful geographical, communications or even entertainment functions. This type of thing has a certain faddish appeal in certain high-tech circles, although it is hard to imagine that it is going to catch on any time soon for most people who live more than 200 miles away from Silicon Valley or its clones in other parts of the world. However, wearable computing may have a great deal of practical use for police and servicemen in the field, whose very lives depend on instant access to computing power. Wearable computing will almost certainly also find important applications for disabled people who, for all intents and purposes, have their natural computing abilities impaired. It is fairly easy to see, for example, how wearable computers could make an important difference to people who are visually and hearing impaired, dyslectic, immobile, and so forth. It is also likely that the emergence of wearable computing will be helpful in the evolution of the next generation of advanced robotics.

There are probably going to be a lot of business opportunities in the wearable computing field in the next decade and some of these will fit the capabilities and resources of firms in the textile industry. But very few of these opportunities are going to have much to do with nanotechnology. There is, however, another area of intelligent fabrics for which nanotechnology will be crucial. These are what might be called "sensor-embedded" fabrics. Sensor-embedded fabrics (my made-up term) are those that respond to the environment in useful or attractive ways. There are a myriad of possibilities. Clothing could, for example, be made to change color as the light changes during the course of the day. This capability could serve the needs of fashion, but also might provide the wearer with warm clothing for cool mornings and evenings and cool clothing for warm days, all in the same item of clothing and without changing clothes! Again, military uses of clothing embedded with sensors may have an important role to serve for detecting biohazards, chemical toxins, and radiation.

Though at some level this kind of functionality could be provided by conventional sensor technology, such applications seem tailor made for nanosensors of various kinds that are embedded in fabrics or for textiles that are intrinsically sensitive to light or some other substance. In Table 6.3, I have summarized the opportunities that seem likely to evolve for nanotechnology in the textile industry. Although, some the first commercial applications for nanotechnology are to be found in the textile industry, some of the opportunities listed in the table are fairly futuristic. In particular, the "smart dust" style nanosensors needed for some of the applications are a few years off—at least at prices that the average buyer of a coat, jacket, skirt, or blouse is likely to pay. The military and police services are almost certain to be the pioneers here.

Generally speaking, my purpose in writing this book is to provide a relatively hard-nosed look at business opportunities that nanotechnology seems likely to bring to various sectors of the economy and not to speculate much on how nanotechnology might help restructure those industries in important ways. I cannot, however, resist the opportunity to note that nanotechnology has the

Table 6.3
Opportunities in Nanotextiles

Opportunity	Comment
Inexpensive "knockoffs" of expensive natural fabrics	Nothing really new here—there have been artificial silks and cottons for decades, but with nanoengineering maybe the new nano knockoffs could be closer to the real thing. But there will be always be people who want to buy natural fabrics, although, if nanofabrics really take hold of the market, silks and satins may become even more expensive than they are now.
Consumer fabrics with enhanced performance characteristics	This is best illustrated at the present time by NanoTex's spill-resistant fabric, but would also include new fabrics that were dramatically stronger, cooler, warmer, and cleaner than any current type of material.
Special-purpose fabric	This is really an extension of the notion of fabrics with enhanced performance characteristics, but for special purposes such as military, police, and fire service use. Essentially, the difference between this and the consumer sector is level of performance. Few people really need their clothes to be bullet-resistant or to stay cool when exposed to a huge fire.
Intelligent clothing	This is clothing that contains built-in electronics, typically sensors that respond to changing environmental conditions. There are many possibilities for types of intelligent clothing of this kind. For example, clothing may change its thermal properties or color in line with atmospheric temperature. There is also a special purpose direction for such intelligent clothing—one could imagine uniforms that contain sensors that warn military, police, and security personnel when toxins and other dangers are around.
Other fabrics	Fabrics used for furniture and curtains would also be the beneficiaries of all of the trends listed above.

capability to change the competitive landscape in the textile industry in some very important ways.

The textile industry has always been one that has gotten chased around the world by low costs. The industry can move quickly from one geographical location to another, as entrepreneurs find new sources of low-cost labor. At various times, textile industries have flourished in the north of England and the northeast region of the United States, only to shift when labor became plentiful in other areas. Of course, the currently favored region for textile manufacture is China. But in time, as the wages of Chinese textile workers rise, this too will change. Perhaps, the textile industry will move to Africa at some time in the (probably not too distant) future.

When the textile industry shifts geographies, it tends to leave economic devastation in its wake. For some reason such problems seem to persist for decades after the change.[92] Poverty and old mill towns seem somehow to go together in songs, stories, and culture. At least they do in the United States and the United Kingdom. Why such poverty persists is not really a question for this book, but presumably the reason that textile firms shift their manufacturing bases so fast is that labor is a very high proportion of total costs, and therefore, it matters what labor costs. As the labor required for manufacturing textiles is relatively low skilled, the textile industry can shift gears, as it were, relatively quickly. Finding textile workers is not exactly like finding skilled programmers.

Now imagine a future, in which a significant proportion of textiles are nano-enabled. Could not this stem the tide somewhat? In such a world, much of the value of textile would be in the intellectual property associated with the textile—or at least much more than it is now. In addition, creating fabrics with embedded nanostructures, such as sensors and the like, might take a higher level of skills and more skilled workers than in the current textile industry, making it harder for firms to change locales. It might also take novel manufacturing processes. For example, consider a sensor-embedded fabric in which the sensors are laid down with an ink-jet printer using a nanometallic ink. Would such a process be easily transferred to an unskilled workforce?[93]

Nanotechnology could therefore alter the course of the textile industry, but the extent of its impact will depend on how much some of the product directions that I have described above catch on. If spill-proof, color-changing, and intelligent clothing prove to be fads or just end up appealing to a small segment of the community, the textile industry will continue to shift around the world to the spot where labor is paid the least. If people in a decade or so come to expect that their clothes should have a growing number of nano-enabled features, then nanotechnology will be creating new opportunities in the textile industry as it transforms the economics of the industry forever.

Whether my speculations on the future economics of the textile industry turn in reality remains to be seen. But it does seem that what nanoengineering will offer the textile industry is an enhanced ability to make fabrics fit the need of the customer, rather than the other way around. This is always an attractive option of offer, of course—in any business. But in the case of textiles, where people spend a lot of time shopping around for the just the right fabric, it is a huge improvement on what we have now. With the rise of digital manufacturing, it is even possible to imagine that people will design their own fabrics and clothes on their desktop PCs, then send the files along to a plant where they will be created and then shipped back to the designer/purchaser. It certainly wouldn't take the rise of nanotechnology to make this possible, but nanotechnology would add a whole new dimension to what kinds of clothes, fabrics, and fashions could be created taking this approach.[94]

Building Nanotechnology

Nanotechnology is likely to find a variety of important applications in the construction industry and many of opportunities for nanotechnology in the building materials segment are really quite similar to those in the textile materials segment. Hear again, we can talk in terms of materials that are designed to waterproof, stain-resistant, and so on. And here again, there is room for sensor-embedded materials that change with environmental conditions. As such, much of what I said about textiles could easily be transferred to the building materials sector with the major exception of my comments on nanotechnology's possible role in changing the cost structure of the industry. (To some extent the building materials industry must remain close to where the construction industry is. Building materials are often quite heavy and become uncompetitive if shipped large distances.)[95]

Rather than make this section simply a repeat performance of the textile section, I want to use building materials as an illustration of how the three major kinds of nanotechnology opportunity that I have described in Chapter 2 actually manifest themselves in practice, since it happens that the building products industry is an excellent exemplar of these different types of new business revenue potential. With this in mind, consider the following three likely trends in building materials as nanotechnology makes its impact felt.

Incremental Nanotechnology Nanomaterials can enhance the physical characteristics of a particular kind of building material. This type of enhancement is a case of *incremental* nanotechnology, since improvements in such characteristics represent a major products direction that the building materials industry has always taken. For example, product managers in the building materials sector have long sought to find new materials for bridges that can take heavier traffic, roofing material that can last longer, and so on. Today such product managers are typically interested in nanotechnology, precisely because it can take them further in their quest. Consider a bridge held in place with carbon nanotube cables—it may be better able to carry traffic and stand high-velocity winds than anything that has gone before.

Evolutionary Nanotechnology An example of intelligent nanomaterials for the building industry would be glass that changes its opaqueness to light in direct proportion to the amount of sunlight to which it is exposed. This is akin to some of the other example of the sensor-embedded technology that I talked about in the section on textiles. It is *evolutionary* technology, in that products of this kind have been around for quite a while, although they are not widely used except in niches. What nanotech- nology can bring to the table is the ability to make intelligent materials more functional and more widely used. Dimensions in which nanotechnology could help include faster response of the intelligent feature, lower-cost, lighter-weight materials, and so on.

Revolutionary Nanotechnology There are surely many different ways that nanotechnology can make revolutionary changes in the building products segment. However, one important direction will certainly be the embedding of various nanoelectronics products into building materials. Some of the nanoelectronics products may take the form of sensors, but their use will have to involve some thing more than could be achieved without nanotechnology. Indeed there is no hard and fast line between evolutionary technology in this sector and revolutionary technology.

For example, high-brightness LEDs (HB-LEDs) are beginning to make their presence felt in the general lighting business, because—at least in theory—they have longer lifetimes, low-power consumption, and brighter lights than more conventional lightbulbs. HB-LEDs are the children of developments in materials technology, especially R&D on the light-emitting properties of gallium nitride. But the arrival of HB-LED technology for use in lights installed in homes, offices, and factories is evolutionary technology. The next stage will be the use of organic LEDs (OLEDs) for lighting systems. Although this is, in a sense, a natural progression (from HB-LEDs to OLEDs, that is), it also means a big switch in materials and production technology—we would now be using organic polymers and perhaps ink-jet printing to create lighting. Nanometallic structures would be used instead of wiring in many instances. But the reason that this is now revolutionary technology is not the technology itself, but rather the opportunity to move in product directions that would not be enabled by either conventional lighting or HB-LEDs. These directions include huge lighting panels that could provide every inch of a wall with a warm glow, an environment that would be quite different from anything that came before. It is even possible that OLEDs could be built into nano-enabled wallpapers and paints, so that lighting would become intrinsic to the wall treatment, again a concept that is highly novel.[96]

Taking things even further, one can imagine nanoelectronics leading to what might be called the "responsive house." This house is widely served with both nanosensors and distributed nonvolatile nanomemories and NEMS devices. It is not just responsive to a particular need but to every possible need of the owners of the house.[97] The movements of people in the house would be monitored by the house system and their preferences would be stored in nanomemories. Responses would be through the lighting, entertainment, heating, and other systems using NEMS and more conventional electromechanical devices. What makes this revolutionary technology is not the individual action of the house control system, but rather its comprehensiveness, that would require certain kinds of electronics that it seems nanoelectronics alone is capable of enabling.

Nothing like what I have described exists today. However, as I pointed out in Chapter 2, *revolutionary* nanotechnology in the sense that we defined it comes with the associated risks. In this case the risks are relatively large. Home management systems have been around for many years and have never really caught on. Is this fault of the particular home management systems being offered and will the enhanced technology that nanotechnology will bring in its wake change the reluctance that consumers apparently have to buy into this kind of thing. The answer is that no one really knows. As it happens, at the time of writing, the consumer electronics industry and the computer industry are very focused on bringing the intelligent home to market, which is the major reason I am using this example. But there are reasons to be skeptical and it should be remembered that the computer industry was talking with much enthuasiasm about the coming of the intelligent home in the early 1980s.[98]

Perhaps a combination of privacy concerns and sheer overkill makes this kind of system a product whose time will never come. But the point that I am trying to make here is not that particular revolutionary changes are going to come about in building materials, but rather that nanotechnology has the potential for propelling this kind of change. With walls that self-light and responsive homes we have come a long way from glass that changes transparency with the weather.

There can be no doubt that nanotechnology certainly holds out the prospect of radical new directions for building materials, although as always nano-enabled products will often have to conform to the norms of consumer demand and industry practice. For example, fashion obviously informs the building materials industry in many ways, although builders are very cautious not to use entirely new building materials, which may fail and lead to complaints or even lawsuits. On the other hand in a few cases, nano-enabled products may be radical and powerful enough to change both fashion—a relatively easy task, or industry norms—a much harder task. It is easy to see, for example, how it could become trendy to use the latest nano-enabled building materials and perhaps the performance characteristics of certain nanomaterials will be so attractive that they will overcome the natural conservatism of the building products industry—they will just be too good not to use.

Nanotech, Food, and Agriculture: A "Nanolithic" Revolution

For some reason agriculture is not often associated with high technology. Yet the Neolithic revolution of some 8,000 to 10,000 years ago, in which farming replaced hunting and gathering as the main way that humans got food, was one of the most profound leaps forward in technology that have ever occurred. Perhaps because the Neolithic revolution occurred so many centuries ago, it seems like farming is not likely to be in tune with the latest technologies.

But the impression that farmers are somehow behind the times is a false one. Farmers, were for example, one of the first industrial groups to adopt online information and e-mail systems, decades before the World Wide Web. Since farms were often located in relatively remote geographical areas this kind of electronic communications enabled farmers to communicate with each other. More importantly it enabled them to gain easy access to long-range pricing and weather forecasts.

I would expect the agricultural sector to be a major early user of nanosensors, too. These can be deployed in a number of interesting ways, which are set out in Table 6.4. One might also add that higher up the value chain nanosensors could also be used in smart packaging to provide information on the freshness or other attributes of a variety of packaged foods.

The use of nanotechnology might also include various nano-enabled sprays and materials to minimize the loss of grains, fruit, and vegetables to disease and inclement weather. Since these materials would be nanoengineered it should be possible to ensure that these remedies would also have fewer bad effects than conventional solutions. In addition, many of the medical breakthroughs due to nanotechnology that I have described earlier in this book would be applicable to livestock. However, the long-term impact of nanotechnology on agriculture has not been widely discussed, although the Foresight Nanotech Institute, has defined "maximizing the productivity of agriculture," as one of six main challenges of nanotech because "Pressure on the world's food sources is

Table 6.4
Uses for Nanosensors in Agriculture and Food Industries

Application	**Description and Comment**
Detection of ripeness of crops	This would be intended to give a more accurate determination of when crops are ripe, which would then improve yields. However, it is not clear whether a nanosensor could be designed that would be better than human instinct in this application. Deployment of such sensors could be in the form of smart dust distributed around the field or orchard. Or a sensor could be deployed in a handheld diagnostic device and used to test individual vegetables, fruits, and so on.
Detection of foodborne pathogens	Use would be in food processing and packing plants and at borders where food is imported. The main objective would be to reduce illness, through more accurate diagnostics. However, the technology would also be labor saving in that fewer food inspectors (government and private) would be needed.
Chemical sensing of fertilizer and insecticides	Nanosensors may be able to optimize the use of fertilizer and insecticide. This will make produce somewhat closer to the organic ideal and also cut costs for farmers.
DNA sensing	Better understanding of plant genetics under different conditions, which will feed back into better strains of food.

ever increasing while harvests have fallen short in recent years. It is anticipated that our world population will swell to 8.9 billion by the year 2050 putting even greater demands on agriculture."

Foresight says that "precision farming, targeted pest management and the creation of high yield crops are a few nanotech solutions."[99] and the Institute clearly hopes that nanotechnology will help to alleviate food shortages or improve the diets of people in the poorest nations. It will certainly help to some extent, and this is also clearly an opportunity for doing well by doing good for some business, but I think that such opportunities should be treated with a tad of skepticism for a couple of reasons. First, the worse cases of long-term poverty and starvation in the world have always tended to be due not to the lack of good food per se, but rather to war and the persistence of corrupt governments. These are not conditions that are easily alleviated by nanotechnology, or indeed any technology.

In fact, agricultural technology has sold best in the most developed companies and is the main reason why farming in North America is so efficient—efficient enough for American farmers to export in large amounts to China, where labor costs are a fraction of what they are in America. It is therefore at least plausible that the latest agricultural nanotech will find a market where previous advanced agricultural technology has found a market, in the First World and not in the Third World. In particular, in the past couple of decades, fresher, more wholesome food has become a major trend in Europe and North America. This is epitomized in the movement towards organic foods, where I believe nanotechnology could play a significant role in this trend, if nanosensors are used to help minimize the use of chemicals and insecticides.

Planes, Trains, and Nanotechnology: Too Many "Opportunities"?

The earlier analysis of how nanotech will impact energy, IT, and electronics, shows quite clearly that nanotechnology will also affect transportation in all its forms. There are just too many ways in which nanotech will bring change to transportation to chronicle them all here. Everything from smart tickets for toll roads to better fuel economy will be promoted, with plenty of products in between. In fact, it is all too easy to get carried away by the variety of opportunities in nano-enabled transport and many popular writers do just that. It does not take long before we are hearing about space hotels and levitating trains, which, it seems fair to say, are not short-term opportunities.

Given the propensity for futurists, engineers, and novelists to dream up new forms of transportation that never make it to reality, I think the impact that nanotech will have on transportation is something that should be viewed with a very critical eye with regard to the extent and timing of the impact. What opportunities can nanotech really find in this sector that firms can start planning for today?

The transportation industry is a sector that doesn't change very fast. The reasons for this are obvious. In the public transportation sector, there are vast amounts for invested capital and even if nanotechnology could make possible true miracles, it is unlikely that the governments and private investors that are involved in the public transport sector would be willing to pull up the railroad tracks and scrap the buses overnight. Similarly, with private sector transportation, the gasoline engine is not going to be gone overnight. There are just too many technological risks for automobile firms to make the leap. There are risks of another sort too, in both the public and private sector, and various nanomaterials are going to have to prove that they are healthy in a number of different ways. Nanocatalysts for fuel will have to be shown to be healthy to passengers, pedestrians, and the environment. And new materials for transport will have to be able to withstand crashes, fires, and so on.

So beware of nanotechnologies selling revenue generating opportunities in the transport sector. That said, there really are some interesting possibilities for the application of nanotechnology to transport that are beginning to emerge now. In Table 6.5, I show where these are to be found and the things to guard against.

Nanotech and Robotics

So far in this chapter we have examined the likely impact of nanotechnology on industry segments that are, and exaggerating only slightly, more or less Victorian in origin. We have seen that nanotech has the ability to actually change some of the basic economic assumptions that have been made in a sector for many years,

Table 6.5

Near-Term Impacts of Nanotechnology on Commercial Transportation

Control	Improved control of vehicles with the use of nano-enabled nonvolatile memories.
Corrosion resistances	Nanocoatings to prevent rust and other forms of corrosion on vehicles. The U.S. Navy is already using nanoparticle coatings on ships.
Responsiveness, comfort, and security	The concept of a smart airframe is already being discussed.This would be embedded with nanosensors to make the plane more responsive to the environment and to security problems. The idea could be extended to car chassis and bodies and to passenger seats for improving comfort. Glass that changes transparency with the amount of light is another idea that falls into this category.
Power	Nanoparticle additives and catalysts for gasoline and diesel. Improved fuel cells. Also nanocoatings of various kinds to improve the lifetime of engines.
Improved materials	Lighter, stronger nanomaterials will reduce costs significant for aircraft and spacecraft. NASA, Boeing, and Airbus are all believed to working on this. The concept is also being employed by the automotive industry.
Payment	Improved payment schemes for public transport and tollbooths with smart cards with nonvolatile memories and sensing devices.

the key example here being the ability of nanotech to add value to the output of textile factories that make them less vulnerable to competition from low labor cost areas.

It is possible that nanotech may also change the outlook for robotics and the related area of artificial intelligence (AI). These two areas have been about to take off "real soon now," for four decades now. Instead they have become niche areas. Robots are widely used in industry, but they are rather dull fellows that can be taught to make a particular kind of car part, certainly nothing like the robot of science fiction, which seems as far off as it ever was. There are also robots that have begun to appear in the consumer market in the past few years, of which the most appealing is surely a robot dog offered by Sony.[100] Similarly AI has survived various fads and fancies about the nature of intelligence and has found its most commercial application in the area of expert systems and neural networks that help make decisions about such weighty matters as loan approvals in the financial services industry and where to drill for oil in the energy industries. Again, as useful as all this may be, it is a far cry from HAL in the movie *2001: A Space Odyssey* (which, for those of you that remember the movie, may not be an entirely bad thing). None of these technologies come anywhere near what people think of intuitively when robots are AI are mentioned in nontechnical discussions. What they are most likely to have in mind is something much more human in at least one of two ways. Robots should look more like human beings (or familiar animals), they will think, and they will have a general purpose intelligence. It would, after all, be hard to come up with a machine that was less like a human than an industrial robot that does one simple task over and over again without tiring and (more importantly) without even the slightest possibility that it will be able to carry out any other action that is not closely related to the one for which it was built.

There are certain trends in nanotechnology that may well take robotics (and perhaps AI) to its next stage. These may present real opportunities for the nanotech sector in the next five years or so. But these opportunities will almost certainly not be of the kind that is often talked about when the overlap between nanotechnology and robotics. Nanorobotics has thus been presented as a Drexlerian vision of tiny self-assembling robots buzzing around reengineering the world. This is the paradise that one finds in some of the Drexlerian literature and the hell that Crichton writes about in *Prey*. It may all be possible some day, but there is almost certainly no money to be made in following this direction for the next couple of decades, except perhaps from writing some best-selling science fiction books about it. And, in any case, this kind of nanorobotics doesn't take robotics any closer to the goal of making robots more like humans or at least like animals.

Where, in my opinion, there is real business potential is in making the kind of robotics we know today a lot better. It seems to me that many of the nano-enabled products and materials that are available today or soon will be are well suited to make big breakthroughs in robotics in terms of the sensitivity, look, mechanics, power and (perhaps) braininess of robots. The reader should understand that what follows is somewhat speculative, but all of it relates to technology that is in the development and commercialization phase and is not just a concept that is being played with by researchers in the academic community.

Sensitivity I have already had much to say about nanosensors and I won't review the material here. Nanosensors would seem to be of considerable potential importance in bringing robots closer to being general purpose machines, closer, in fact, to the human paradigm. We have numerous sensors of many different types in our bodies and while there may be different ways in which someone might use a general purpose robot, ranging from the inspiring to the downright frivolous, it would seem that a great leap forward in reengineering human senses in a robotic environment are essential to all of them.

In the robot of the future sensors would have to be deployed that replicate the five animal senses. Smell and taste would be replicated with gas and chemical sensors—probably more than one kind would be used.[101] Touch and hearing would be recreated with pressure and motion sensors, and vision with optical sensors. Of course, all of these sensors can be created and used in robots without any need for nanotechnology. So what is so special about nanosensors? As I have mentioned, nanosensors are potentially much more sensitive than other kinds of sensors and (again potentially) could be manufactured in low-cost arrays that can be embedded throughout the "skin" of the robot, so that it will be much more fine-tuned to its environment than is now possible.

Look Speaking of skin, there have been many attempts to create a human-looking robot and there have been some real achievements made in the effort, in the sense that some robots that have been built on a one-off basis really look like people. The problem, of course, is finding a way that will create robot flesh and robot eyes that can be easily and economically reproduced in such a way that robots can be made in some volume.

At the present time, there are really no technologies that are easily scalable in this way. However, some of the artificial skin technologies that I have discussed previously in the section on regenerative medicine would seem to have a second line of applications in robotics. Of course, not all robots of the future will have to look like humans or familiar animals. There are types of robots where this will not be necessary—there is no particular reason why robots whose purpose is to work in dangerous environments (power plants, under the oceans, etc.) should look like humans, although *all* robots will probably benefit from a certain level of biomimicry, including nano-enabled biomimicry. But I suspect that robots that serve the function of pet or a helpmate will benefit a lot from looking like a human, dog, cat, or horse.

Mechanics and Mobility One of the ways in which today's industrial robots differ from humans is that, for the most part, they don't move around. There *are* a limited number of mobile robots, but these are either toys or are used for very specialized purposes, as for defusing bombs, for example. Most industrial robots are simply clever but fixed machine tools.

Nanotechnology could add to the mobility of robots in two ways. First, it could supply new sources of long-lasting power—we are probably talking fuel cells here, but it is possible to imagine how better batteries or photovoltaics might play a role. Second, with the introduction of NEMS devices, robots can be built that can respond in more subtle ways and offer a broad range of physical responses. Back in the 1980s, when practical robotics was first coming into existence, getting robots to move in anything other than the clumsiest ways was a huge challenge. Although things are better now, NEMS could produce a great leap forward in this area.

Braininess There is already a huge amount of literature on just how intelligent an AI system can really be and widely divergent opinions on whether AI can truly match human intelligence. These are interesting debates, but ones that have little to do with the matter at hand.

Back in the 1960s, artificial intelligence research centered around the notion that human being possessed a sort of general intelligence that could be replicated if only processors were fast enough. The problem is that they weren't anywhere fast enough. AI eventually changed directions, in part because the generalized intelligence idea was leading nowhere and in part because there was recognition that the human brain wasn't based on a single very high-speed processor, but rather on a distributed network of quite slow processors. This insight led to the development of neural networks. Nonetheless, there is little doubt that the development of faster processors is key to realizing the notion of "strong AI," the idea that AI can reach and even surpass the level of human beings.[102] And as we have seen in the chapter on nanotechnology and electronics, one way to faster processors lies through nanoelectronics.

Summary: Key Takeaways from This Chapter

This chapter covers several relatively unrelated industries, but I hope it gives some sense of how pervasive nanotechnology can be and how broad the opportunities are likely to be for businesspeople. Unlike the chapters that covered electronics/semiconductors, energy, and the life sciences, this chapter was more focused on market developments than technological specifics. Some important things to remember from the chapter are the following.

1. Although most of the work on complex "nanoproducts" is somehow related to the electronics/semiconductor, energy, or life sciences industries, there are many other sectors of the economy on which nanotechnology will make an impact. Often these are just extensions of the products being developed in the three "big three areas" with which this book is primarily concerned.
2. The chemicals and materials industries seems well positioned to take immediate benefit from the nanotech revolution. A few firms in this sector are being very proactive about nanotechnology, but most are sitting by and watching what happens until nanotech shows some real commercial promise. Once nanotech's promise is fulfilled, I believe that there will be an excellent fit between nanotech and the chemicals and materials sector, with the possibility that the materials and chemicals sectors may be largely driven and structured around developments in nanotechnology.
3. The textile industry also has much to gain from nanotechnology. For many years, it has been pulled around the globe to areas where the wages are lowest and it has been hard to establish technology-based areas of competitive advantage which might have kept the industry in more developed regions. Nanotech seems to hold out the possibility of creating relatively protectable IP and other kinds of value added in the textile sector that could change the economics of the industry. In terms of business opportunities in this segment, these are considerable. Spill-resistant, nanoengineered textiles are already available and are usually cited as one of the first commercial nanotechnology products. But there are much more elaborate implications of nanotechnology for the textile industry in the form of intelligent fabrics of various kinds.
4. The impact of nanotechnology on the building products sector will be similar in many ways to its impact on the textiles sector. It will also be fairly broad based and will include many opportunities for

evolutionary nanotechnology to make materials that are stronger, better insulated and so on—in other words, taking building materials to the next stage by following the design trends that have always been with us. But more radical developments are also possible that may be compelling enough to overcome the natural reluctance of builders to use untested materials. Nanomaterials may even become fashionable for building in certain circles.

5. In agriculture, most of the applications for nanotechnology seem likely to involve nanosensors, which will show, for example, when food is fresh and ready to pick. Another product direction lies in materials that protect food from disease and insects. Despite the popular image, farmers in the West have always been open to new technologies, although one of the hopes by some nanotechnologists is that nanotechnology will also help to alleviate food shortages in the Third World. It remains to be seen what the future geography of agricultural nanotechnology will look like.
6. In transportation, there are a myriad of applications for nanotechnology. These include opportunities for new materials to make the transport lighter, stronger, more fire retardant, and so on. It also includes improved power systems and more comfortable and safe public transport.
7. The current state of robotics is a long way from that envisioned by science fiction writers, where industrial robots are specialized self-learning machines and consumer robots are little more than toys. The usual association between robotics and nanotechnology is in terms of self-assembling nanorobots that are designed to reengineer their environment. Another different way that nanotechnology may impact robotics is through making machines more general purpose and human-like in terms of sensitivity, braininess, look, and mobility.

Further Reading

There are few, if any, books on nanotech that deal with the topics covered in this chapter. I would recommend scanning current issues of trade magazines in each of the sectors (and in any others in which you may be interested) for the occasional story on nanotech. Typically, however, the coverage will not be especially well informed, because the writer will be well versed in knowledge about the sector, but not about nanotech.

A useful resource in understanding what the impact of nanotech is likely to be in various sectors of the economy is the Web site[103] and the publications of the Foresight Nanotech Institute. Originally founded by Eric Drexler, Foresight remade itself in the under the excellent leadership of Scott Mize. The Institute is now reinventing itself as the place to go for information on nanotech and its impact on society. Two other useful sources are *Converging Technologies for Improving Human Performance: Nanotechnology, Biotechnology, Information Technology and Cognitive Science*, which is edited by Mihail C. Roco and Williams Sims Bainbridge[104] and *Five Regions of the Future: Preparing Your Business for Tomorrow's Technology Revolution.*[105] Both books deal with topics much broader than just nanotechnology, but both are good and serious source books on where technology is headed and it is an interesting exercise to fill in the gaps and define just where nanoengineering will play a useful commercial role.

7

How to Conduct a Nanotech Implications Audit in Your Company

Why an Audit?

You have reached the final chapter. In what has come before I have tried to show that the emergence of practical nanotechnology means new, exciting and (most importantly) profitable opportunities in many sectors of the economy, especially those related to electronics, energy, and the life sciences. As I have also noted, the opportunities are especially noteworthy when they coincide with major megatrends, such as the need for better mobile communications, looking after the needs of aging baby boomers and solving the so-called energy crisis. However, the commercial impact of nanotechnology is broader than just the opportunities it presents.

Nanotechnology also presents firms with threats. It may be that if your firm does not take advantage of nanotechnology to redesign your products for better customer acceptance, your competitors will. Much the same thing can be said about the use of new nanomaterials or (perhaps) of the latest nanotools. There is also the issue of how best to adapt an organization to the arrival of commercial nanotechnology. Do you need new kinds of technical staff? Or some kind of retraining program to deal with new issues of safety and marketing?

With the nanotechnology business about to take off, now is clearly the time for both large and small firms to stop talking about how cool the technology is, forget the hackneyed quips about the next big thing being very very small and come up with some real business plans.

Making the transition from techno-speculation and PR-babble to real planning can be quite scary. Understanding value chains, building business models is all a very inexact science, and mistakes can cost millions. As an industry analyst who covered the rise and fall of the telecom industry, I am convinced that the critical input to business plans for products based on novel technologies is *realistic* market forecasting and planning. During the optical boom, I saw firms swarming with brilliant minds talk themselves into believing that the markets they were chasing after were worth billions of dollars, when they were worth millions. They built business models based on this assumption, and then they crashed and burned.

Had they built those business models around the truth, many of these firms would be with us today. There are other firms out there who would have survived if only they had not underestimated the new networking technology. Most notably these include retail firms that could not adapt to the Internet era. This might have been because they thought that the Internet was no more than high-tech hype only to discover far too late that their customers had all moved to the Web. It might have been because they tried to reengineer their business for the age of the Web, but were simply incompetent in doing so.

Ten years from now it will be possible to tell similar tales about firms that either overestimated or underestimated the coming nanotech era. This is why it is so important to carry out an "audit" or your organization now to see just where nanotechnology will have an impact. This need is heightened by the fact that the impact of nanotech is likely to be quite broad, because, as I noted earlier in this book, nanotech is not exactly an industry, but more an enabling (or platform) technology that will make itself felt in almost every sector. The audit need not necessarily be a complex process, but it does need to be a comprehensive and rational one. The goals are to see how the impact of nanotechnology can impact costs and revenues, and, hence, profits, over a period of time.

It is impossible in a book of this kind to give specific examples of how such an audit (call it a "nanotech implications audit") should be performed. Few firms have yet performed such an audit. That is one reason I am providing this guide in the first place. In any case, every business is going to be impacted in quite a different way by nanotech, so it seems to make sense, at least at this stage of nanotech development, to sketch out a general approach and leave it at that.

There will no doubt be changes in this methodology in later editions of this book, but at this stage of the game I will confine myself to describing what I see as the essentials of such an audit. Since the impact of nanotechnology on many business is likely to be both diverse and far ranging, the rules for the audit will have much to do with setting limits on which areas, issues and so on that need to be considered in the audit. Some aspects of the audit process described below are essentially identical with any management consulting or auditing procedure. Others are more specific to nanotechnology.

The Nanotechnology Impact Analysis in Summary: A Six-Step Program

There seem to me to be six important steps that need to be taken in any nanotechnology impact audit. These steps that I see as making up the nanotech implications audit process are as follows:

- Establishing objectives;
- Fact and trend selection;
- Establishing a data collection methodology;
- Opportunity analysis;
- Threat analysis;
- Organizational analysis.

Establishing Objectives

This step consists of deciding why you are carrying out the audit in the first place. While such an audit may be extremely useful for many types of organization, each is going to have its own special set of concerns and (perhaps) differing time frames.

Audits of organizations are usually performed with some specific audience in mind. Sometimes this audience is going to be external investors, as in the case of the regular financial audit carried out on publicly traded companies, by public accountants. Sometimes, audits are performed to inform the concerns and expectations of senior management. This second type of audit, I believe, is going to characteristic of most nanotech implications audits for the foreseeable future.

With this in mind, the first thing that needs to be done is to interview the senior management staff for whom the audit is designed to understand what they would regard as a successful outcome to the audit, and that is, what they are really looking for out of it. They may be focusing on opportunities, threats, required organizational changes, or the impact on customers, suppliers, or competitors. In any consulting process it always makes sense to know in advance what the client wants, and the nanotech implications audit is no different in that regard.

There is an important caveat, however. As every seasoned consultant has discovered, clients do not always know precisely what they want and therefore the audit objectives should not be set too narrowly. If it is clear that the primary concern of senior management is the deleterious impact that nanotech may have on the business, this *does* mean that threat analysis must be the main focus of the audit. It *does not* mean that if other important issues and items are uncovered during the audit, they should be ignored. For example, if a major opportunity for the business is unearthed, this is clearly worth mentioning, if only as way of putting the threats into better perspective.

Another important aspect of objective setting is considering time frames. While a nanotech implications audit may take into considerations that are short-term (a term that economists usually take to mean one year or under) or very long term (more than a decade), I am convinced that a five-year time frame for an audit makes most sense. This is because the very short-term time frame is unlikely to turn up many important implications that are not obvious before the audit was commenced. The very long-term audit becomes highly speculative and is most likely to seriously *under*estimate the impact of nanotechnology.[106] In addition, most companies do not have serious planning horizons that go beyond just a few years. Also, and this is the basis for most of what follows, it is easy to take a reasonable stab at which particular nanotechnologies are going to impact your business in a five-year period. Over long periods it almost becomes anybody's guess. However, I acknowledge that some organizations will want to conduct audits that extend their reach over longer periods than just five years. I believe that they can use the principals set out in this chapter to achieve such an audit, but the further out they push the audit, the more they should also interpret the results with caution.

Fact and Trend Selection

Much of the first step in the nanotechnology implications audit is little more than common sense and really doesn't have that much to do with nanotechnology per se. By contrast, selecting the issues, facts, and trends that will be used as the raw data for your nanotech implications audit involves nanotech specifically.

Nanotechnology may be very broad-ranging, but only some of it has relevance to your business. This step consists in picking the data that is most relevant to your particular organization and the objective of the audit. This book will have prepared you for this stage, because the medium-term impact of complex nanotechnology is likely to be felt through a relatively small group of products—the ones that we have focused on most strongly in this book.

The starting point for this stage should be your organization's core activity. For small organizations, this is a relatively trivial task, since the core business is probably a single product or product line. For large businesses that have hundreds or thousands of products it will be necessary to simplify, since the analysis to be performed becomes simply too complex to carry out. One way to simplify is to apply the well-tested 80/20 rule which says that 20 percent of one's products (or product lines) account for 80 percent of one's profits. In the audit and analysis, focus on that 20 percent.

With this accomplished the next part of the analysis is actually determining which trends within nanotech are important for the future of your organization. Some of this analysis may be patently obvious. If you are in the oil business, the relevance of nanocatalysts designed to improve the efficiency of gasoline hardly needs any further explanation. However, simply operating at this obvious level is really missing the point of the audit in the first place, which is to provide a bigger picture of the implications that include some implications that are less than obvious.

This does not mean that the obvious should be ignored. But it does mean that all the important ways in which nanotech could impact the business need to be taken into account. But again, in order to accomplish this, a considerable amount of simplification is required. It is simply impossible to consider every possible angle in which nanotech can impact a business or other organization. This impact may come from new nanoengineered materials or from more complex nano products. Since the focus of this book is on complex nano products, we will only consider these here. But a very similar process could, and should, be performed with nanomaterials as its subject, where this seems relevant.

For complex nano products, I believe that the initial simplification should be to consider first the impact of nanoelectronics, nanoengineered energy product,s and nano-enabled products in medicine and pharmaceuticals. As I have stated throughout this book, this is where most of the impact of nanotechnology is going to come from in the next five to ten years. Having said that, some of these areas may be dismissed fairly quickly as having little or no impact on a particular business. If you are in the cardboard carton business, you may well have an interest in how nanoelectronics will lead to smart packaging solutions, but you may have relatively little interest in the nanoengineered life medical solutions, for example.

It is possible to simplify even further. As we have already seen, only some areas within these three major areas of nanotech are going to have significant commercial potential in the next few years. For example, although nanoelectronics may ultimately have numerous applications, what is really going to matter in terms of market impact in the near future are nanoengineered computer memories, sensors, and displays. Similarly, within the life sciences, what will really matter will be drug discovery, drug delivery, and regenerative medicine. As long as the time frame is a relatively short-term one, these areas are all that need to be considered to get a pretty clear idea of what nanotechnology will mean to your organization. In addition, because the impact of nanotechnology is often felt through the key "megatrends" discussed earlier in this book, it makes considerable sense to also consider the impact of these megatrends on your business and how nanotechnology factors into this, something which has we have also discussed in some detail.

Finally, do make sure that in selecting the areas that you will focus on, you take into consideration not just the items that are related directly to the products and services that you offer, but also ones that will impact the inputs to your business, especially materials and components and also those that might change the competitive landscape in your industry.

Establishing a Data Collection Methodology

Having decided what data and trends are important for the audit, the next stage is to consider where the data that you need to consider will come from. The likelihood is that it will come from both internal and external sources.

Internal sources include interviews with your own sales and marketing and technical staff (perhaps your financial people too). Internal sources of this kind are something of a double-edged sword. On the one hand you are all part of the same organization, so there is some reason to think that such people will be more open about providing the required data. On the other hand the obvious impacts of office politics cannot be ignored in this process. Where internal sources are involved, therefore, it is an important part of the audit to make sure that your organization is well prepared to provide the information that you need. In particular, it is vital that staff that are supposed to be providing information understand that the audit has the backing of senior management.

Internal sources—marketers, sales people, product managers, and engineers—are often quite close to the market, so they are in a good position to understand what is really going on. But also always remember that is their job to do marketing, sales, product management, or engineering. It is almost never to collect data.[107] So the data collected from internal sources is suspect because those sources themselves will have gathered it in the most casual manner. Anecdotes and conventional wisdom can be useful at times, but it can hardly serve alone as the most important input to the nanotechnology implications audit and, in any case, internal sources of data are always going to be biased towards the views of the most senior level of management in the organization being audited. In one context, this is a good thing, as, at some level, everyone needs to be on the same page to make the organization work. However, every so often a "sanity check" on the collective beliefs of an organization can be useful.

This sanity check means going to external sources. These may include trade publications, specialist market research studies, attending conferences, or doing online research. Each has its own function and an analysis of the role of each of these external sources lies outside the scope of this book. However, Table 7.1 provides a brief guide to how each of these sources can help in the nanotech implications audit and what its limits are. The reader should also note that in the context of the audit, each source comes in two flavors: the pure nanotech flavor and the sector flavor. By this I mean, for example, that a reader interested in finding out more about nanotech in the energy industry may chose to attend a

Table 7.1
A Brief Guide to External Sources for the Nanotechnology Impact Audit

	Why Use It?	**Limitations**
Trade press and newsletters	Timely information from diverse primary sources. Inexpensive.	Often written by journalists with limited knowledge, especially of *both* nanotech and sector-specific information. Articles on nanotech that appear in trade magazines for specific sectors are frequently quite naive about nanotech.
Published market research studies	Provides much greater depth than any trade press article could ever provide and with better economic analysis. Usually based on a broad range of expert opinion. Excellent sanity check for internal sources of information.	Expensive. Analysts do not typically have lengthy industry experiences and so may ignore some the practices common in an industry sector. Poor market research studies either simply repeat received wisdom or what everyone in the industry already knows.
Conferences and trade shows	Provides access to views of industry insiders and the opportunities to meet and discuss issues with peers and competitors.	Presentations and presentation materials at conferences are often poor, limiting what can be gained by attending. Presentations are sometimes thinly disguised advertisements for the firms that make them.

conference of nanotechnology that provides significant coverage of energy issues or an energy industry conference that provides a panel or two on nanoengineered energy solutions. If time and money permits it, attending both kinds of conference is likely to be the best possible strategy, since each is likely to have a different window on the world and come at the problem of nanoengineered energy sources, transport, and storage from different directions.

One final word on data collection: whether you are considering internal or external sources, it is vital that sources be both broad and deep. In terms of depth, the people that you talk with must be well informed. This much is obvious. But it is also important to talk to people across the board. For example, the view of the CTO will be quite different from that of the COO. Be aware that there is often a trade-off between depth and breadth. That CTO may be able to tell you more than you need to know about how nanotech will impact the various technological subsystems used in the organization, but his or her view on how it will impact the day-to-day workings of the organization may be as uninformed as anyone else.

Opportunity Analysis

This step consists in identifying where the key nano-related data that you have collected in the previous steps will impact your organization in terms of both generating new revenues and cutting costs.

As I mentioned above, there is a temptation to think of this process as applying only to the final product of your organization, but it is also important to think about opportunities as potentially being located in your supply chain and in your organization itself. For example, likely improvements in IT brought on through nano-enabled pervasive computing solutions may provide efficiencies for firms in working with their suppliers. And nanoengineered energy solutions may lower costs within your organization.

Thus for each nanotechnology of importance, it is important to think through the opportunities that will arise in products, organization, and supply chain and, as I have pointed out there are, in fact, relatively few nanotechnologies that are likely to have profound impact on businesses and other organizations within the time frame within which it seems reasonable to perform nanotechnology impact audits. These observations lead directly to the opportunity analysis itself. In Table 7.2, I bring together what has come before to perform a somewhat limited and generic opportunity analysis of the kind I have in mind here. The real thing would be a lot more specific to the needs of the person/organization performing it and the cells of the matrix would be filled with a lot more ways that nanotech will have an impact.

Table 7.2
A Generic Nanotechnology Impact Audit Opportunity Analysis

	Products	**Organization**	**Supply Chain**
Nanoelectronics			
Displays	Improved quality and lower cost of existing products sold with displays. Adding displays, including flexible displays and e-paper, to products or packaging.	Better display-enabled communications to improve organizational efficiency.	Display-enabled smart packaging. Enhancements to advertising displays.
Sensors	Nanosensors added to products for enhanced security, product safety, etc. More responsive products including robotic systems.	Improved building security and health at work.	Sensor-enabled smart packaging. Improved inventory and tracking systems.
Memories	Adding intelligence and storage capabilities to products of all kinds.	Improved IT facilities and ease of storing corporate data. Better mobile communications.	Improved IT and tracking systems for supply chain management.
Nanomedicine			
Drug discovery and delivery	While only applicable to certain sectors, some of the spinoffs from this area may have impact for chemical analysis and delivery of small amounts of materials in other sectors.	Improved health for employees from better drug delivery systems and new drugs.	
Regenerative medicine	Applications primarily limited to the medical sector, but may find other uses in robotics.	Fewer sick days, and better and more immediate treatment for on-the-job injuries.	
Energy			
Photovoltaics	Low-cost power for small devices of all kinds, new kinds of building and engineering materials that provide power as one of several features.	Improved power sources for mobile communications and computing devices. Improved environmental protection.	Similar to organization, as a whole but improved tracking and inventory systems using solar power handhelds will be a notable implication.

Table 7.2 (Continued)

Fuel cells	Long-lasting backup power and systems for recharging batteries. Some applications for outdoor and portable power. Longer-term applications in transport.	Potential reduction in energy costs. Improved power sources for some portable equipment.	Potential reduction in freight costs.
Electricity transport and storage	Opportunities to sell services based on remote/networked energy sources.	Electricity cost reductions.	Reduction in freight costs to customers and from suppliers.
Gas additives	Novel petroleum-based products.	More fuel-efficient heat systems and transport.	More fuel-efficient transport.

In both my generic example and in the real world the cells in the matrix are filled in through a process of "brainstorming." For understanding how the various nanotechnologies will impact products, you will have to ask and answer a few simple questions such as: Where will those cool nano-enabled features really matter to customers? How much will they matter? Can these features be used to create entirely new species of product? Can we apply nanoengineering better than our competitors? And how are the answers to all these questions likely to change over time? Similar questions can be asked of the impact on the supply chain. Could nanoengineering speed up deliveries, for example, or help to reduce inventories? And they can be asked with regard to the impact on the organization itself.

Once again, let me state that the table above is highly generic and a real-world analysis would be a lot more detailed and, frankly, a lot more interesting, than what I show here. And while the table above represents the "bare bones" of an opportunity analysis of this kind, there are a number of ways in which this analysis can be made more comprehensive and potentially more helpful to senior management:

- *Group brainstorming.* While the "brainstorming" process can be performed successfully by one person, it is possible to use a research panel instead. The advantage of a group approach is that expertise that no one person could possibly possess can be brought to the table. The disadvantage is that the more extreme views are filtered out. This may sound like a good thing, but "extreme" may also sometimes mean the same thing as *original.*

- *Quantifying the impact.* The analysis in the table is purely a qualitative one. However, it is possible to quantify the impact in various ways. For example, if you are looking at the impact on your organization of nano-enabled energy technologies you might want to think through what the cost effect would be on your products, organization and supply chain of these technologies being implemented. If energy— at least the kind of energy with which the particular nanotechnology innovation you are considering is concerned—is a small part of your costs, then the "nano-impact," may be marginal. On the other hand, if you are in the airline business, a nanocatalyst that increases the efficiency of fuel by (say) 20 percent, will transform your business.

Threat Analysis

This step will aid in the discovery of the dangers that lurk in nanotechnology. The entire process is merely the converse of what we have described for the threat analysis and most—if not all—of the same methodological comments apply.

One aspect of threat analysis is to consider what it would mean to you if your competitors seized the opportunities presented by nanotechnology before you did. In fact, the answer may not be all that obvious. If your competitors steal your thunder in this way, of course, it may mean lost revenues for you. But then again, there are plenty of examples in business history of firms being second into a market and still being the winner. In addition, there may be no reason to suppose that you can't continue to pull significant business from areas that are not enabled in any meaningful way for many years to come. For example, as I have mentioned several times now, nano-enabled computer memories are clearly a big market opportunity that could be worth tens of millions of dollars in a few years. But there will still be plenty of room for firms that make plain old DRAMs or even EPROMs for many years into the future.

Analyzing threats from nanotechnology to your organization and supply chain is somewhat harder to do than to your product. This is because the natural tendency of all but the most nanophobic is to think of nanotechnology as improving things, not making them worse. As with products themselves, one possible threat is that your competitor will adapt organizationally to nanotechnology before you. (Indeed this could be a bigger threat than your competitor using nanotech to come up with new products.) But the issues are bigger than that and a little unpredictable. Questions that you should ask yourself include:

- Would using nanomaterials in your organizations raise new health and safety issues?
- Will new kinds of training be required for nano-enabled products?
- Are you ready to take on the additional technology risk associated with adopting a product?

Organizational Analysis

This step is designed to pinpoint the changes that must occur to make sure that your organization makes the best of the opportunities and prepares the best possible defenses for the threats facing it from nanotechnology. In other words, the two steps that preceded this are theoretical in nature, while this step is a practical one.

Precisely for that reason it is hard to provide generic guidance for this step, which is about tactics, not strategy. The kinds of things that we are talking about here include some level of training in the basics of nanotechnology for general management and, in a sense, it includes the nanotech implications audit itself. One particularly important issue to consider is whether the audit, training and other related matters should be a "one-off," or whether it should be carried out on an ongoing basis. The answer to this question probably depends on just how important nanotechnology is to the organization. At this point it seems that nanotechnology is likely to become pervasive enough over the next few years that almost every organization should consider its implications. Some kind of audit such as I have suggested above is therefore recommended for almost any kind of business and perhaps some training too.

That said, some organizations are going to find that nanotechnology will have very limited impact on their business for the foreseeable business in the future, at least in a direct sense and where management can make a difference. However, at the other end of the scale, there are businesses that are deliberately choosing to adopt nanotechnology as a key part of their strategies and value propositions. For such firms it will certainly be necessary to keep track of developments in nanotech on an ongoing basis and provide frequent updates to management on a regular basis. It may also make sense under these circumstances to construct alternative scenarios for the business under different assumptions about how important nanotechnology. It is typical in scenario planning to consider three different sets of assumptions that are in some way high, low, and medium growth. In this case, one might consider the impact on the business of a world in which nanotechnology is adopted very quickly, one in which it gradually penetrates important sectors and one in which nanotech turns out to be more hype than anything else.[108]

Summary: Key Takeaways from This Chapter

There are two key takeaways from this chapter, and in a sense from this whole book:

1. As the importance of nanotechnology grows it will become imperative for organizations of all types to assess the impact that nanotechnology will have on both the organization's bottom line and on the organization itself.
2. The process of assessing this impact—the nanotechnology impact audit—should follow a well-defined procedure, such as the one I have set out in this chapter. This audit should include various steps comprised of establishing objectives, fact and trend selection, establishing a data collection methodology, opportunity analysis, threat analysis, and organizational preparation. In an organization for which nanotech is likely to be an important influence, the audit, training in nanotech, and perhaps related scenario planning should become an important part of the management process.

Further Reading

As far as I am aware there are no other books that have discussed anything similar to the nanotechnology implications audit. However, many of the ideas in this chapter have been influenced by Michael Porter's great classic, *Competitive Strategy*,[109] which I believe would prove a useful tool to anyone working on a nanotechnology implications audit in that it does an excellent job of tracing the relationships between the various kinds of firms and internal groups active in a business ecosystem. These relationships can each be explored with regard to the change that nanotech may be able to create in how they operate.

In addition to Porter's work, there are probably many other good books on the process of auditing and organizational development that will be of considerable use in carrying out such an audit, but none stands out as essential reading. A browse of the management and accounting sections of a large bookstore may well prove rewarding in this area.

Appendix: Eleven Essential Information Sources for Nanobusiness

My original conception of this book included a long appendix listing as many business-related sources for nanotech that I could think of or find. However, as the book evolved, I decided to include a brief bibliography/"websiteography" for each of the chapters. In this appendix, however, I am going to list what I see as the top eleven sources of information that everyone in the nanotech business should read or consult regularly.

Before presenting my list, let me also provide a few caveats.

First, this list is highly subjective, personal, and perhaps a little eccentric. Other people might have provided a different list, although I would wager that there would be quite some overlap between all such lists offered by reasonably knowledgeable observers of nanotechnology. That said, some of these "reasonably knowledgeable observers," would certainly object to omissions on my list and they would be right to do so, in the sense that there are certainly some popular books and URLs that are not included below. Nonetheless, in following this field, I have come to believe that many of the popular nanotech Web sites are little more than homes for reprints of the latest nanotech-related press releases or softball interviews with industry luminaries.

My second caveat is a reiteration: this is a list of sources for *business* opportunities. Many of the sources that I list are not technical in any useful sense. You will not learn that much of use for engineering a nanotech product from the sources listed below and the awful truth is that I would not be that surprised if you found a few technical inaccuracies in some of the sources I cite below. This, I believe, does not invalidate them in any significant way for the purpose of setting strategy and keeping up with "nanobusiness."

It was also my intention originally to include some discussion of the best trade shows and conferences to attend to find out what is going on in the world of nanobusiness. However, at the moment, some of the big nanoconferences in the United States seem to be going through a period of transition and repositioning. At the same time, the world of nanoconferences seems to be fragmenting with specialist conferences on nanomaterials, nanomedicine, and so on becoming more prominent. From a purely personal point of view the most valuable trade show/conference that I have attended is the annual event organized by the Nanotech Science and Technology Institute (NSTI),[110] which at the time of writing alternated its venue between Boston and Anaheim. I believe this is also the best-attended U.S. nano event. The best-attended conference/trade show in the world is almost certainly the major Nano Tech show held in Tokyo.

Finally, while what follows is a "top ten," the list below is not meant to imply any specific ranking. Each source has its own unique characteristics and to compare them all would take us rapidly into the realm of comparing apples and oranges.

A.1 *The Diamond Age*

There is now a complete science fiction subgenre in which nanotechnology is central. Most of these books have no conceivable literary or scientific merit and most are based on a kind of crude Drexlerian approach to nanotech. There are a few exceptions to this blanket condemnation. Michael Crichton's *Prey* is one such exception. *Prey* is a great read, but it has very little to do with the type of nanotech that is likely to be commercialized in the near future and it is confined to a very particular kind of nano product—ultrasmart microscopic nanomachines, which are a long way off commercially, although that is not the way that Crichton presents them. In any case, nobody would claim that *Prey* was a good source of business ideas.

The Diamond Age by Neal Stephenson[111] is different. Like *Prey*, it goes well beyond what is now possible, but in describing an entire world dominated by nanotechnology in the way that the West in the 19th and early 20th centuries were dominated by steam, it is an excellent source of ideas of product directions that nanotechnology might take us. It also contains some interesting ideas about the way that the abundance created by nanotechnology could transform society, which might also be thought provoking for the would-be nano marketer.

Of course, *The Diamond Age* is a novel, not a marketing tool, and should be read as such. However, even if the book is read primarily as a novel, its complex plot is extremely entertaining and the book should probably be read a couple of times to be really appreciated. In any case, the widespread familiarity of the nanotech community with this book probably makes at least one read essential to anyone who really wants to be a nanotech insider, even if he or she doesn't want to treat the ideas in this book especially seriously.

A.2 *The Singularity Is Near*[112]

This book is another great source of ideas. It was written by Ray Kurzweil, the famous inventor and techno-optimist. Like Stephenson in *The Diamond Age,* Kurzweil has painted a picture of where nano- technology may be taking us. However, this is a work of nonfiction, not fiction.

Nanotechnology is only part of what this book is about. The basic theme is to investigate the notion of the "singularity," which was an idea that the science fiction writer Vernor Vinge dreamed up a couple of decades ago. The singularity is a point in time in which humanity has been transformed utterly by the arrival of advanced technology to the point at which they are no longer quite the same species. Humans will become immortal, able to travel to the stars, and so on. What makes all this interesting from the perspective of Kurzweil and people like him is that he believes that the singularity is going to occur quite soon. The reason he believes this is essentially that he views technology as evolving on an exponential curve that starts out slowly and then accelerates to an unbelievable pace. Not much changed technologically in the 3,000 years that ancient Egyptian culture was dominant, but by the 20th century, there seemed to be a major new technology appearing every few years or so. What Kurzweil does is take this to its logical conclusion.

What has any of this have to do with nanotechnology? Kurzweil thinks that nanotechnology is one of three disciplines that will help bring about the singularity. The others are genetics and robotics. (Kurzweil talks a lot about GNR, which is an acronym for genetics, nanotechnology, and robotics.) Much space is devoted to nanotechnology and the reader cannot help but take away some ideas that will be useful in business development, product planning, and so on. This is particularly the case, because, although Kurzweil has a rather Drexlerian cut on nanotech, he comes at issues with the eye of the practical inventor and entrepreneur, which is likely to make this book more readable for a businessperson than would a similar book from an academic or a journalist. That said, Kurzweil's view of the future tends to be not just *trans*human but also inhuman—the sort of future that only a technology nerd could love. (Stephenson's fictional world is much more believable.) It may therefore be that the best way of dealing with this 652-page book for the average reader will be dipping into it, not reading it cover to cover.

A.3 *Nanotechnology: Basic Science and Emerging Technologies*[113]

Most business people looking for a primer on nanotechnology will find their way to the book by the Rattners that I cited in Chapter 1. This is an easy and informative read, but is ultimately fairly superficial and any businessperson who is getting seriously involved with nanobusiness is going to need more.

My suggestion for such a person would be *Nanotechnology: Basic Science and Emerging Technologies*, which was put together by a team of Australian authors and is published by Chapman and Hall. This is quite comprehensive and, although the reader comes away with a fairly deep knowledge of nanoscience and nanotechnology, the book never strays from high-school level science. In addition, the book does a reasonably good job of relating technology developments to real world applications and devotes chapters to nanotools, nanopowders and nanomaterials, carbon nanotubes, molecular manufacturing, nanobiometrics, nanophotonics, nanoelectronics, and "future applications."

A.4 Nanotechnology.com

When I was an analyst covering the optical networking business, there was one Web site that was essential reading. It provided all the news that was fit to print on optical networking, as well as powerful analytical articles, product reviews, financial news, and so on.

No one has yet come up with anything like this site for the nanotech sector, even though there are now more Web sites devoted to news about nanotechnology than you can shake a stick at. Nonetheless, nanotechnology.com stands out as being of particular interest to business people. The is perhaps because the man behind then site, Darrell Brookstein is a financial expert and the author of his own fine book, *Nanotech Fortunes*, which focuses on all aspects of stock market investment in nanotech.

At the time of writing, nanotechnology.com was still a work in progress. However, a visit to www.nanotechnology.com gives a good idea of where the site is headed. Here you will find a nano blog (the only promising one since Howard Lovy's blog—see below—went bye-bye) and separate parts of the site devoted to financial, government, and business matters. There will also be the usual calendar and a job section. Unlike some other similar sites, there is clearly a serious commitment from Brookstein and his firm, the Nanotech Company LLC, to transforming this site into *the* site that the nanotech community will have to turn on a daily basis in the not-too-distant future.

A.5 Nanotechweb.org

This site is run by the (U.K.) Institute of Physics (IOP) and is also an excellent source of information on things nanotechnological. As one might expect from the Institute of Physics, this is a different kind of site from nanotechnology.com and is much more technically oriented.

Nanotechweb.org is probably the single best source of information on technical developments in nanotechnology. A regular review of this Web site will keep the busy corporate executive and entrepreneur up to the minute on what is going on in the industrial labs and universities. While much of this information appears in other places, most of the other sites are nowhere near as comprehensive in their coverage. I should also mention that the IOP also publishes a technical journal called (naturally enough) *Nanotechnology*, about which (again naturally enough) you will find more information at nanotechweb.org. However, *Nanotechnology*, is intended for researchers, academics, and Ph.D.-level students, and most of the papers are far too detailed to be of real interest to the average businessperson.

A.6 NanoMarkets

NanoMarkets LC is the firm that I cofounded and for which I work. Our primary goal is to provide the nanotechnology sector with reliable business information, mainly in the form of reports which combine market research and detailed market projections. Our coverage is mainly in the electronics and semiconductor space, but we also examine the markets for nano-enabled products in the energy and medical segments. We are one of the few firms of our kind to focus exclusively on nano-enabled markets in this way. Our reports are sold both to giant materials, chemical, pharmaceutical, and electronics firms, and to smaller entrepreneurial organizations. It is these reports that have provided much of the material that you will find in this book.

NanoMarkets' Web site provides information about the reports that we sell, but we also think of it as a resource for business-oriented nanotech-related information. Much of this is free for the taking. The freebies include a variety of white papers on topics such as "Six Opportunities in Nano-Enabled Drug Delivery Systems," and a "Roadmap for Printable Electronics." There are a growing number of shorter pieces that we call "Perspectives," which can be downloaded directly from the site; the URL of which is http://www.nanomarkets.net.

A.7 Nanoinfo.jp

Clearly NanoMarkets is not the only source of market research in the nano space. What the world needs is a single site that brings this all together. They now have this in the form of http://www.nanoinfo.jp, which is a spinoff from the world's largest reseller of industrial market research reports. It is possible to order reports directly from this Web site.

Unfortunately, at the present time, this is really a Japanese Web site, although most of the reports listed are described in English. There is also some free content in English.

A.8 The Foresight Nanotech Institute

As discussed elsewhere in this book, the Foresight Nanotech Institute (formerly the Foresight Institute) has gone through some important transformations in the recent past. It is now *the* think tank to go to for nanotech issues and, I suspect, likely to become quite influential in the future. It is no longer simply the vehicle for Eric Drexler's thoughts and seems to have accepted a broader definition of nanotechnology, although "molecular manufacturing" is still central to its concerns.

In the same way that www.nanotechnology.com is the best site to go to for business information and www.nanotechweb.org is the best URL for technical information, Foresight's Web site (http://www.foresight.org) is the best site for thinking on the social implications of nanotech, although the more libertarian among us may not be completely comfortable with some of the Institute's approaches. At the time of writing, Foresight has framed its goals in terms of a number of challenges where it believes that nanotech can make a significant difference. These challenges include the big issues: global energy, clean water, health and longevity, productivity of agriculture, IT everywhere, and development of space. It is also sponsoring a "nanotechnology roadmap initiative," in conjunction with Battelle, a leading R&D organization.

In addition to the information found on its Web site (which includes some very useful links), the Foresight Institute is a membership organization, which anyone seriously interested in nanotechnology should become a member of. And Foresight runs one of the best annual nanotech conferences, which is, again, a must-attend for professional and managers who find themselves focused more and more on nanotechnology matters.

A.9 *Seeing What's Next: Using the Theories of Innovation to Predict Industry Change*[114]

This is the latest book from Clayton Christensen (along with two coauthors), and if the index is to be believed, it does not mention nanotechnology even once. Despite this—and the fact that this book is a very difficult read—I believe it is an essential read for everyone involved in nanobusiness.

Christensen has provided a clear meaning for the much abused term, "disruptive technology," which is something that nanotechnology is frequently claimed to be. In this book, he and his coauthors go into a great deal of detail about the changes to expect from various kinds of innovation as they impact various industry sectors. At the very least, this book is an antidote against the tendency to assume that everything coming out of the nanotech labs is "disruptive." At best it is an excellent guide to forecasting the implications of particular nanotechnologies. The case studies provided on the future of education, aviation, semiconductors, healthcare, and telecommunications are very helpful in this regard.

A.10 *Nanotechnology Law and Business*

This journal describes itself as the "first authoritative source of information entirely devoted to the legal, business, and policy aspects of nanotechnology and more generally small scale technologies." It is actually the *only* source of information "entirely devoted to the legal, business, and policy aspects of nanotechnology," for although *Scientific American, Technology Review,* and *Red Herring* now give regular coverage to nanobusiness, this still represents a fairly small part of their output.

NLB is published both in hardcover and on the Web, but is most accessible on the Web at http://pubs.nanolabweb.com, where you can buy a subscription. Issues covered include markets, intellectual property, regulatory matters, and finance.

A.11 *Nano-Hype: The Truth Behind the Nanotechnology Buzz*[115]

From the name, one might expect this book to be a diatribe against the overhyping of nanotech. And there is actually some of this. However, the book is really a critical appraisal of the reality behind nanotechnology and is one of the best guides I have come across to the people and institutions that are building and contributing to the nanotech sector.

Coverage includes profiles of leading entrepreneurs, investors, and government officials active in the nanotech field and an excellent review of activities beyond American shores. The author, David M. Berube, is the research director of Nanoscience and Technology Studies at the USC NanoCenter.

In memoriam. If my book had been written just a few months earlier, one Web site that would have definitely appeared on my list would have been Howard Lovy's Nanobot blog. Very sadly this was discontinued when Howard went to work for a nanotech firm. However, the entire blog can still be found, "frozen in time," as Howard puts it, at http://www.nanobot.blogspot.com. At the time of this writing, there seemed a reasonable chance that Nanobot will eventually be revived in some form.

Endnotes

1. Kurzweil, R., and T. Grossman, *Fantastic Voyage*, Emmaus, PA: Rodale Books, 2004.
2. I am using "telecommunications" here in a broad sense. It includes both the activities of telephone and cable companies along with networks put in place within businesses. I also consider it to cover everything from the Internet to fiber optic and wireless communications.
3. Some people in the investment community were apparently expecting a nanotech boom to follow Nanosys' proposed IPO. In the end this IPO was withdrawn, which only served to generate more skepticism about commercial nanotech.
4. Drexler, K. E., *Engines of Creation*, Garden City, NY: Anchor Press/Doubleday, 1986
5. Khosla's comments were made at the MIT/Stanford/UC Berkeley Nanotechnology Forum in 2004 and widely reported in the investment press at the time.
6. Varchaver, N., "Is Nanotech Ready for Its Closeup," *Fortune Magazine*, May 17, 2004 (Web edition).
7. Popper, K. R., *Logic of Scientific Discovery*, London and New York: Routledge, 2002 (originally published in German in 1935).
8. I am thinking of the late Professor Richard Smalley's accusation that Eric Drexler's futuristic view of nanotechnology would scare children.
9. See http://www.nano.gov/html/facts/whatIsNano.html.
10. Ratner, M., and D. Ratner, *Nanotechnology: A Gentle Introduction to the Next Big Idea*, Upper Saddle River, NJ: Pearson Education, 2003, p. 7.
11. A particularly good one is J. Al-Khalili, *Quantum: A Guide for the Perplexed*, London: Wiedenfield & Nicolson, 2004.
12. Wilson, M., *et al., Nanotechnology: Basic Science and Emerging Technologies*, Boca Raton, FL: Chapman & Hall/CRC Press, 2002, p. 58.

13. Wilson, M., *et al., Nanotechnology: Basic Science and Emerging Technologies*, Boca Raton, FL: Chapman & Hall/CRC Press, 2002, p. 81.

14. Aerogels are constructed using the sol-gel manufacturing process and are three-dimensional networks of nanoparticles with air or some other gas trapped in the network. Aerogels are porous, extremely lightweight, and extremely strong.

15. According to Wilson, M., et al. in endnote 13, "Zirconia . . . a hard, brittle ceramic, has even been rendered superplastic by nanocrystalline grains."

16. Ziegler, J., and H. Puchne, *SER—History, Trends and Challenges*, Cypress, p. ix.

17. Ziegler, J., and H. Puchne, *SER—History, Trends and Challenges*, Cypress, p. 3-3.

18. Schmidt, R. R., and B.D. Notohardiono, "High-End Server Low-Temperature Cooling," *IBM Journal of Research and Development*, Vol. 46, No. 6, 2002; http://www.research.ibm.com/journal/rd/466/schmidt.html.

19. It is not clear whether Hero actually invented the concept of a steam engine or whether he was simply describing an idea that had been around for a while.

20. Ratner, M., and D. Ratner, *Nanotechnology: A Gentle Introduction to the Next Big Idea*, Upper Saddle River, NJ: Pearson Education, 2003, p. 13.

21. The talk was published in 1960 as Feynman R. P., "There's Plenty of Room at the Bottom," in the February 1960 issue of California Institute of Technology. It is reproduced in full at many Web sites dealing with nanotechnology and related subjects. See, for example, http://www.its.caltech.edu/~feynman/plenty.html.

22. Sometimes the invention of the term "nanotechnology" is credited to Drexler, who certainly talked about it in *Engines of Creation.* Sometimes the term is ascribed to Norio Taniguchi, a Japanese engineer who used it in technical writings published in 1974.

23. *The Diamond Age* is a wonderful science fiction novel by Neil Stephenson set in a world dominated by Drexlerian nanotechnology. This nanotechnology is, as far as one can tell, largely carbon-based, hence the reference to diamond.

24. Regis, E., "The Incredible Shrinking Man," *Wired Magazine*, October 2004 (Web edition).

25. It is important to remember that the objection to Drexler's views being raised here are not that they are just "too far out," but rather that they are inconsistent with the laws of nature. Even anti-Drexlerians can be visionaries. While Drexler and his friends campaign tirelessly for support for molecular manufacturing, anti-Drexlerian Smalley equally campaigns tirelessly for a national energy policy built around nanoengineered electricity storage and distribution systems (more on this in Chapter 4).

26. Formerly the Foresight Institute.

27. Regis, E., "The Incredible Shrinking Man," *Wired Magazine*, October 2004 (Web edition), op cit.

28. Ratner, M., and D. Ratner, *Nanotechnology: A Gentle Introduction to the Next Big Idea*, Upper Saddle River, NJ: Pearson Education, 2003, p. 43.

29. Chou, S.Y., "Nanoimprint Lithography," in *Alternative Lithography: Unleashing the Potentials of Nanotechnology*, New York: Kluwer Academic, 2003.

30. See endnote 10 for reference.

31. See endnote 12 for reference.

32. See endnote 4 for reference.

33. See endnote 24 for reference.

34. Storrs Hall, J., *Nanofuture: What's Next for Nanotechnology*, New York: Prometheus Books, 2005.

35. Atkinson, W. I., *Nanocosm: Nanotechnology and the Big Changes Coming from the Inconceivably Small*, New York: American Management Association, 2003.

36. Uldrich, J., and D. Newberry, *The Next Big Thing Is Really Small: How Nanotechnology Will Change the Future of Your Business*, New York: Crown Business, 2003.

37. Sotomayor Torres, C. M., *Alternative Lithography: Unleashing the Potentials of Nanotechnology*, New York: Kluwer Academic, 2003.

38. Brookstein, D., *Nanotech Fortunes: Make Yours in the Boom,* San Diego: The Nanotech Company, 2005.

39. See endnote 36 for reference.

40. This is now being asserted as a real possibility by respectable scientists, most notably Ray Kurzweil.

41. My estimate would be that the capital necessary for a nanomaterials firm capable of volume production would be measured in the tens of millions of dollars. This would be well within the capability of many firms financed by venture capitalists. Even in volume production, the volumes involved would be very small compared to the bulk chemical industry.

42. MEMS devices, as I have already mentioned, are often built in older fabs and have significantly larger features than leading edge electronic chips that may be categorized as genuine products of nanotechnology.

43. This is Texas Instrument's for high quality displays. DLP stands for digital light processing. For more on DLP, see http://www.dlp.com.

44. The sense of disruptive used here is a little different than that of Clayton Christiansen in his now-classic books, where he seems to mean a technology that can tap into a pent-up demand because of its low cost.

45. Lux Research press release describing a study jointly by Lux and the patent attorney firm Foley & Lardner.

46. *Discover* magazine, special "Frontiers of Science" issue, October 2005, p.59.

47. Fishman, T. C., *China, Inc: How the Rise of the Next Superpower Challenges America and the World*, New York: Scribner, p. 217.

48. See, for example, Crews, C. W., "Nanotech's Choice: Pork or Innovation," *Monthly Planet*, Competitive Enterprise Institute, August, 2004.

49. Gasman, L. D., "Telecompetition: The Free Market Road to the Information Superhighway," Cato Institute, Washington D. C., 2004.

50. McKibben, B., *Enough: Staying Human in an Engineered Age,* New York: Henry Holt, 2003.

51. "IBM Exec: Impending Death of Moore's Law Calls for Software Development Changes," *InfoWorld*, May 24, 2005.

52. None of which should be taken to mean that the old ways of doing microelectronics will vanish from this earth. Old fabrication plants will find new uses for novel (non-nano)

devices and mature processing and memory technologies. Consider the history of PROM memory, still widely available and used, but quite technologically obsolete.

53. Although Moore's Law has been assumed in any and every forecast of electronics related market sectors, it can sometimes lead to quite surprising developments. Back in the early 1990s, I attended a conference in Washington, D.C., on the topic of gigabit networking. At the conference I heard many different views of the potential for networking at speeds of 1 Gbps and above. The pessimists thought that it would never be possible to run the Internet's TCP/IP protocols at speeds as high as 1 Gbps; no one thinks that there are any practical limits to TCP/IP anymore. The optimists thought that there was a good chance that all the major Internet backbones would run at 1 Gbps by the turn of the century. As it turned out, many major public backbones now run at 10 Gbps and 1 Gbps connections to PCs are becoming quite common. I don't think anyone at the conference would have found this a believable scenario. This is a good illustration of how easy it is to underestimate how far technology can develop in a decade, something to remember when thinking about where nanoelectronics may take us.

54. That the ITRS is overly conservative is a point that has been made to me in many interviews with semiconductor industry executives. It is hardly surprising that this is the case, since consensus forecasts seldom come up with dramatic predictions. For the latest output of the ITRS, see http://public.itrs.net/.

55. This means that it must be able to use existing manufacturing infrastructure and also interface easily with CMOS devices.

56. By that time Moore's Law will have taken leading edge commercial devices to a point where they will have considerably outpaced anything that traditional silicon microelectronics can build. At this point the only options for the semiconductor industry will be to adopt nanoelectronics wholesale or move in entirely new directions.

57. MWNTs are a bit like Russian dolls—one nanotube inside another.

58. See Gasman, L., "Carbon Nanotube Electronics: A Technology Analysis and Market Forecast," May 2005, NanoMarkets.

59. For more details see the IEEE's press release, "IEEE Study Group Begins Work on Quality Standard for Carbon Nanotubes," January 7, 2005, which can be found at http://standards.ieee.org/announcements/pr_icnqtsg.html. The IEEE has always been very proactive in developing standard for new kind of electronics and its rapid embracing of new kinds of electronics should be heartily applauded. However, precisely because it is willing to take on early stage technologies, some of the IEEE's standards around certain technologies are now no more than memories as the result of those technologies failures in the marketplace.

60. See Gasman, L.,"Magnetic Memory: An Analysis and Forecast of the Market MRAM," January 2005, Richmond, NanoMarkets.

61. The Nobel Prize in chemistry was awarded to Alan Heeger (USA), Alan MacDiarmid (United States and New Zealand), and Hideki Shirakawa (Japan) for "the discovery and development of conductive polymers."

62. See discussion of catalysts in Chapter 4 of this book for more on why this makes a difference.

63. For our purposes here we consider printable and plastic electronics as part of the same general area. However, as we have noted in the main text, not all printable electronics is

based on polymer inks—metallic inks are also possible. In addition, not all plastic electronics uses a printing platform for manufacturing. Thus the $400 million OLED market is largely composed of products produced with deposition techniques, not printing.

64. Ink-jet printing of circuits is one of the first real-world examples of digital manufacturing, a topic that is being pioneered at MIT and promises to bring a "desktop" manufacturing revolution similar to that wrought in computing by the PC. For more details of this trend see "Fab," by Neil Gerschenfeild.

65. An ASIC is a circuit designed for a specific purpose, rather than use an off-the-shelf memory or logic chip. The advantage of taking the ASIC route is that ASICs provide higher performance for the applications for which they were designed. But they are expensive to make, especially if they are manufactured using a process that needs a mask.

66. For an excellent discussion of all the many applications of quantum dots, see McCarthy, Wil, *Hacking Matter: Levitating Chairs Quantum Mirages and the Infinite Weirdness of Programmable Atoms*, New York: Basic Books, 2003.

67. Indium phosphide is still an important material for building photonic devices, but has not quite lived up to its original promise, which to some extent was based on the huge supposed need for optical devices during the telecom boom era. In addition, indium phosphide has proved harder to work with than some people expected.

68. The editors of Scientific American, *Understanding Supercomputing*, New York: Scientific American, 2002.

69. Luryi, S., J. Xu, and A. Zaslavsky, *Future Trends in Microelectronics: The Nano Millenium*, New York: Wiley-IEEE Press, 2002.

70. Goser, K., P. Glosekotter, and J. Dienstuhl, *Nanoelectronics and Nanosystems from Transistors to Molecular and Quantum Devices*, New York: Springer, 2004.

71. Turton, R., *The Quantum Dot: A Journey into the Future of Microlectronics*, New York: Oxford University Press, 1995.

72. Huber, P., and M. Mills, *The Bottomless Well: The Twilight of Fuel, the Virtue of Waste, and Why We Will Never Run Out of Energy*, New York: Basic Books, 2005.

73. A recently built military housing complex in Hawaii is using PV as its primary heat source. The management has guaranteed tenants low-cost housing that includes all utility costs. Under this regime it becomes very important that management's costs are as predictable as possible and PV is able to provide much more certainty in this regard than fossil fuels.

74. For more on nanocatalysts, see "NanoWorld: Nanocatalysts for Oil, Drugs," Charles Choi, Washington Times, March 26, 2005.

75. "NanoWorld: Nanocatalysts for Oil, Drugs," Charles Choi, Washington Times, March 26, 2005.

76. "Quantum Dot Materials Can Reduce Heat, Boost Electrical Outputs," NREL News Release, May 23, 2005.

77. Editors of Consumer Reports, *Consumer Reports Electronics Buying Guide*, New York: Consumers Union of the United States, 2006, p. 32.

78. In the United States at least, there are fast-track regulatory mechanisms for drugs that show especial potential for diseases that are likely to prove fatal over short periods of time. Regulation of drug introduction continues to be a subject for ongoing debate, with pro-regulatory forces claiming that the public is not being adequately protected and the

more practical among us pointing to the number of people who die waiting for drugs to be approved.

79. See endnote 1 for reference.

80. See, for example, Christensen, C. M., *The Innovator's Dilemma,* New York: Harper Collins, 1997.

81. One should recall the price that Faust was asked to pay.

82. Clark, A., *Natural-Born Cyborgs: Minds, Technologies and the Future of Human Intelligence,* New York: Oxford University Press, 2004.

83. Freitas, R. A., "Nanomedicine: Volume I: Basic Capabilities," *Landes Bioscience,* 1999, and "Nanomedicine: Volume IIA: Biocompatability," *Landes Biosciences,* 2003.

84. Goodsell, D., *Bionanotechnology: Lessons from Nature,* Hoboken, N. J.: Wiley-Liss, 2004.

85. Niemeyer, C. M., and C. A. Merkin, *Nanobiotechnology: Concepts, Applications and Perspectives,* New York: John Wiley, 2004.

86. Knowingly quoted out of context, but nonetheless, an irresistible quote for the beginning of this chapter.

87. This also makes nanotechnology hard to analyze and pinpoint from a commercial point of view.

88. The efforts to standardize carbon nanotubes by the IEEE and others will also hasten this commoditization process.

89. Bulk nanomaterials are almost certainly not a short-term market opportunity. However, if nanotechnology becomes a truly epoch making technology—a core underpinning of our whole society, a bulk nanomaterials business would almost certainly emerge. The best depiction of such a society to date is found in Neal Stephenson's novel, *The Diamond Age.*

90. It should be stressed that nanotech doesn't really need a Netscape to justify itself as a sector and a nanotech boom will probably not include some of the downright silliness associated with the dot.com boom and bust, if only because the barriers to entry for nanotech are sufficiently high that nanotech firms will have to have thought through their business models better than some the Internet businesses or a decade or so back.

91. The history of the plastics industry, which in many ways is so similar to the emerging nanotech industry, provides many examples here—consider bakelite, nylon, and rayon, for example.

92. I attended the University of Manchester in England in the late 1960s, decades after the textile industry had collapsed in the area. I was struck immediately by how much poverty there was compared to my native London, just a few hundred miles away. It was really not until the 1980s that Manchester completely sloughed off the legacy of the decline of textiles.

93. My comments on this matter should be taken as a general comment on the economics of the textile industry as illustrated by its history over many years. At the present time, Chinese dominance in this industry is a function of an ultralow cost, but reasonably skilled labor force. The low costs reflect fixed currency rates and the recent history of China. China's competitive edge is likely to dull somewhat in future years, as the Chinese economy matures.

94. At the time of writing, digital manufacturing has become something of a hot topic in certain circles with much of the interest centered around the work of MIT professor, Neil

Gerschenfeld and his book *Fab*. Digital manufacturing has little direct connection with nanotechnology, but they complement each other beautifully and both are part of wider revolution in production and materials technology.

95. It occurs to me that by it ability to make building materials lighter, nanotechnology might actually help enable building material manufacturers reach a larger geographical market. Whether this could ever turn into major industrial trend remains to be seen.

96. OLEDs transform energy into light and it is not surprising therefore that a very similar technology to OLEDs, but in reverse could serve as a photovoltaic cell,transforming light into energy. It would be equally possible to imagine photovoltaic arrays being encapsulated in wall treatments in the same way that we describe for OLEDs in the main text.

97. We could just as easily be talking about an office.

98. *The Economist* magazine recently published an article about the wave of new products that were aimed at the intelligent home and some good reasons to suppose that this market would not take off. In this context, the reader should remember that Apple's famous television ad, showing how personal computing would create the automated home, was shown in 1984.

99. http://www.foresight.org/challenges/index.html.

100. In the 1980s there was a sudden fad for personal robotics that lasted a few years before fading out, once people realized that these robots would never do anything very interesting. It seems to me that the current generation of personal robotics is only slightly more impressive than what was on the market in 1984.

101. There is a growing trend to create nanosensors that can detect a broad range of chemicals/gases. But there is a limit to what can be done within costs and size limits.

102. For more on strong AI, see the works of Raymond Kurzweil or Hans Moravec.

103. http://www.foresight.org.

104. Roco, M. C., and W. S. Bainbridge (eds.), *Converging Technologies for Improving Human Performance*, Boston and London: Dordrecht, 2003.

105. Barker, J. A., and S. W. Erickson, *Five Regions of the Future: Preparing Your Business for Tomorrow's Technology Revolution*, New York: Portfolio, 2005.

106. As I have noted elsewhere in this book, there is a strong tendency for forecasters to overestimate the effects of new technologies in the short term and underestimate them in the long term.

107. Some firms maintain their own internal market research units. However, these units frequently rely on outside sources for their own information.

108. See Peter Schwartz, *The Art of the Long View*, which has become the classic work on scenario planning. The reader should note that in a later book, Schwartz discusses nanotech's impact on business and other types, although he does so rather briefly.

109. Porter, M., *Competitive Strategy: Techniques for Analyzing Industries and Competitors*, New York: The Free Press, 1980.

110. http://www.nsti.org.

111. Stephenson, N., *The Diamond Age: Or, a Young Lady's Illustrated Primer*, New York: Spectra, 1995.

112. Kurzweil, R., *The Singularity is Near: When Humans Transcend Biology*, New York: Viking Adult, 2005.

113. See endnote 12 for reference.

114. Christensen, C. M., *Seeing What's Next: Using Theories of Innovation to Predict Industry Change,* Cambridge, MA: Harvard Business School, 2004.

115 Berube, D. M., *Nano-Hype: The Truth Behind the Nanotechnology Buzz,* Amherst, N. Y., Prometheus Books, 2006.

About the Author

Lawrence Gasman is the principal analyst at NanoMarkets, LC., a market research and consulting firm specializing in the emerging technology sectors that are being empowered by developments in nanotechnology and other developments in materials science.

Mr. Gasman has more than 25 years of experience of analyzing the commercialization potential of complex technologies and in the recent past has conducted demanding market research on organic electronics, printable electronics, displays, nano-enabled computer memories, nanosensors, thin film photovoltaics, smart packaging, photonic integration, MEMS, network processors, and semiconductor lasers. His has advised some of the largest electronics, semiconductor, and telecommunications firms in the world, as well as high-tech start-ups, venture capitalists, and investment banks.

Mr. Gasman has been quoted a wide range of publications, including *The Wall Street Journal*, *Investor's Business Daily*, *Business 2.0*, *Red Herring*, and *Small Times*, and has been a frequent speaker at major nanotechnology and telecommunications conferences. His three previous books have dealt with telecommunications technology and policy and he has testified to Congress on the future of the FCC. He is a member of the IEEE, a senior associate of the Foresight Nanotech Institute (where he serves on the editorial board), and a senior fellow at the Cato Institute, a leading Washington, D.C., think tank.

Mr. Gasman holds advanced degrees in scientific method and business administration from London University, as well as a bachelor's degree in mathematics from the University of Manchester. He lives in Crozet, Virginia, and can be reached at lawrence@nanomarkets.net.

Index

Recent Related Artech House Titles

Nanotechnology Applications and Markets, Lawrence Gasman

Organic and Inorganic Nanostructures, Alexi Nabok

Optics of Quantum Dots and Wires, Garnett W. Bryant and Glenn S. Soloman

Semiconductor Nanostructures for Optoelectronic Applications, Todd Steiner, editor

Advances in Silicon Carbide Processing and Applications, Stephen E. Saddow and Anant Agarwal, editors

Microfluidics for Biotechnology, Jean Berthier and Pascal Silberzan

Fundamentals and Applications of Microfluidics, Nam-Trung Nguyen and Steven T. Wereley

Mathematical Handbook for Electrical Engineers, Sergey A. Leonov and Alexander I. Leonov

Nanotechnology Applications and Markets, Lawrence Gasman

Nanoelectronics: Principles and Devices, Mircea Dragoman and Daniela Dragoman